Urban Traffic Pollu

Also available from E & FN Spon

Transport Titles:
Transport Policy and the Environment
Edited by D. Bannister

Transport and Urban Development
D. Bannister

Transport Planning
In the UK, USA and Europe
D. Bannister

Transport, the Environment and Sustainable Development
Edited by D. Bannister and K. Button

Urban Public Transport Today
B. Simpson

Pollution Titles:
Water Quality Assessments, 2nd edition
A guide to the use of biota, sediments and water
in environmental monitoring
Edited by D. Chapman

Water Pollution Control
A guide to the use of water quality management principles
R. Helmer and I. Hespanhol

Water Quality Monitoring
A practical guide to the design and implementation of
freshwater quality studies and monitoring programmes
Edited by J. Bartram and R. Ballance

Toxic Cyanobacteria in Water
A guide to public health consequences and their management
in water resources and their supplies
Edited by I. Khorus and J. Bartram

A Water Quality Assessment of the Former Soviet Union
Edited by V. Kimstach, M. Meybeck and E. Baroudy

Urban Traffic Pollution

Edited by
Dietrich Schwela and Olivier Zali

E & FN Spon
An imprint of Routledge

London and New York

This edition published 1999
by E & FN Spon, an imprint of Routledge
11 New Fetter Lane, London EC4P 4EE

Simultaneously published in the USA and Canada
by Routledge
29 West 35th Street, New York, NY 10001

Printed and bound in Great Britain by
TJ International Ltd, Padstow, Cornwall

Publisher's Note
This book has been prepared from camera-ready copy provided by the editors.

British Library Cataloguing in Publication Data
A catalogue record for this book is available
from the British Library

Library of Congress Cataloging in Publication Data
A catalog record for this book has been requested

ISBN 0 419 23720 8

TABLE OF CONTENTS

FOREWORD

Many urban areas of this world have high concentrations of air pollution sources resulting from human activities; sources such as motor vehicle traffic, power generation, residential heating and industry. Urban air pollution not only represents a threat to human health and the urban environment, but it can also contribute to serious regional and global atmospheric pollution problems. Air pollution is experienced in most urban areas and is therefore a world-wide problem and an issue of global concern. It has been estimated by WHO that globally about 500,000 people die prematurely each year as a consequence of exposure to ambient pollution of suspended particulate matter. Increases in morbidity from respiratory diseases due to air pollution are estimated to occur in about 40 million people; several million infants die each year from acute respiratory infections exacerbated by air pollutants.

The marked increase in urban populations occurring in many cities, together with industrialisation, will lead to an increase in the emissions of pollutants and to an increase in public and environmental exposure to these pollutants. By the year 2000, close to 6,000 million people will be living in the world, and it is expected that about 45 per cent of them will be in urban areas.

Motor vehicles are now recognised as the major contributor to urban air pollution; their emissions of suspended particulate matter including sulphuric acid aerosols, lead, carbon monoxide, nitrogen oxides and photochemically reactive hydrocarbons affect the health of populations. Given this situation, a pure description of urban traffic pollution in terms of monitoring results is not adequate; the underlying problems have to be tackled by taking action for the implementation of clean air. In almost all of the large cities of the world, air pollution and noise from motor vehicles are, or are fast becoming, major problems for the physical and mental health of the people. Industrialised countries, where 86 per cent of the world's vehicles are to be found, have extensive and long-standing experience of the problem. In developing countries, rapid industrial growth and population increase, together with rising standards of living will probably lead to patterns of motorisation that resemble those of industrialised countries. Since the 1960s, the number of motor vehicles in the world has been growing faster than its population.

Air quality standards are often exceeded during many days of the year in the majority of cities, especially in developing countries. If cities continue to grow at their current rate, air pollution, and consequently effects on health, will worsen unless forceful measures are implemented. Urgent steps are needed in heavily polluted areas in many cities of the world and particularly in regions of Eastern European, Eastern Mediterranean and Western Pacific countries, as well as in Latin American and South East Asian cities. In these cities, measures are required to control industrial emissions and to prevent air pollution from road traffic by the development and implementation of comprehensive transport policies and urban planning, and by phasing out leaded petrol and introducing obligatory emission control in cars.

The Republic and Canton of Geneva has developed a comprehensive and sophisticated air pollution control programme designed to meet Swiss Federal air quality standards relating to pollutants derived from motor vehicle emissions. The Canton of Geneva has taken a strong lead in tackling the problems of air and noise pollution in the Canton in a very efficient way. In recognition of the importance of the problem of motor vehicle air pollution world-wide, WHO and the Ecotoxicology Service of the Department of Public Health of Geneva together supported an earlier WHO document, *Motor Vehicle Air Pollution. Public Health Impact and Control Measures* (WHO/PEP/92.4), published in 1992, and which formed the foundation for the present book.

This present book begins with a general introduction to the issues of motor vehicle air pollution in Chapter 1. Chapter 2 provides a comprehensive review of the health problems associated with emissions from motor vehicles, while Chapter 3 deals specifically with the health effects related to motor vehicle noise. Chapter 4 describes how people are exposed to motor vehicle air pollutants and gives an estimate of the numbers of people who are exposed to them in traffic, alongside busy roads and in residential areas of high traffic density. Control of motor vehicle pollution is reviewed in Chapter 5 by considering the efforts made to reduce emissions and to reduce vehicle usage. Case studies from various cities around the world are presented in Chapter 6 to show how various control actions have been applied to reduce air pollution from motor vehicles. A detailed report of motor vehicle pollution and its control in Geneva (Switzerland) is presented in Chapter 7 in order to provide a broad outline of the different kinds of activities that must be considered so as to evaluate air pollution in urban areas, to estimate future growth and to develop a staged sequence of control measures acceptable to the public and economically feasible in the given situation.

We hope that this book will provide essential information and encouragement to all countries in their efforts to deal with the problems created by motorisation. The value of this book lies in its usefulness for municipal authorities in air quality planning and management. Therefore, the target audience of this book is the scientific community, staff from municipalities working in the field of air and noise pollution abatement and non-governmental organisations which are interested in improving the air and noise situation in the communities.

Municipalities can adopt and adapt the ideas and experiences from the case studies presented here for their particular situation. Through the sharing of this experience, cities may be able to avoid mistakes made in the past and to introduce effective measures in the near future to reduce or limit damage that has already been incurred. Many cities will need to begin planning or applying more strictly a progressive motor vehicle emission control strategy that is feasible and affordable and that will alleviate the immediate air pollution problem and reduce human suffering from respiratory diseases.

World Health Organization

Republic and Canton of Geneva

Contributors

Mr François Cupelin
Service Cantonal l'Ecotoxicologie
Case postale 78
1211 Genève 8, Switzerland
Tel. +41 22 781 01 03, Fax +41 22 320 67 25
E-mail <françois.cupelin@ecotox.etat-ge.ch>

Professor Peter G. Flachsbart
Department of Urban and Regional Planning
University of Hawaii at Manoa
Porteus Hall 107
2424 Maile Way
Honolulu, Hawaii 96822
Tel. 1 808 956 7381, Fax +1 808 956 6870

Dr David T. Mage
United States Environmental
Protection Agency USEPA
MD 56
Research Triangle Park, NC 27711 – USA
Tel. +1 919 541 1327, Fax +1 919 541 1486

Dr Isabelle Romieu
c/o Dr Mauricio Hernández
Rollings School of Public Health
Dept. of Occupational &
Environmental Health
1518 Clifton Road
Atlanta, Georgia 30322 – USA
Tel. +1 770 488 76 40 Fax +1 770 488 73 35
E-mail <iar9@cdc.gov>

Professor Ragnar Rylander
Department of Environmental Medicine
University of Gothenburg
Medicinaregatan 16
413 90 Gothenburg
Sweden
Tel. +46 31 773 3601, Fax +46 31 825004
E-mail <ragnar.rylander@envmed.gu.se>

Dr Dietrich Schwela
Urban Environmental Health
Division of Operational Support in
Environmental Health
World Health Organization
20 Av. Appia
CH-1211 Geneva 27, Switzerland
Tel. +41 22 791 4261, Fax +41 22 791 4127
E-mail <schwelad@who.ch>

Mr Ashwani Singh
APEX Consulting Services
Rue de Genève 148 bis
CH-1226 Thônex, Switzerland
Tel. +41 22 860 13 66, Fax +41 22 860 13 69
E-mail <info@apexserv.com>

Mr Michael P. Walsh, WHO Consultant
3105 N. Dinwiddie Street
Arlington, VA 22207 – USA
Tel. +1 703 241 1297, Fax +1 703 241 1418

Dr Olivier Zali
Service du Chimiste Cantonal
Quai Ernest-Ansermet 22
CP 166
1211 Geneva 4, Switzerland
Tel. +41 22 328 75 11, Fax +41 22 328 0150

ACKNOWLEDGEMENTS

The co-sponsoring organisations, WHO and the Republic and Canton of Geneva, wish to express their appreciation to all those whose efforts made the preparation of this book possible. An international group of authors contributed, and for some chapters more than one author and their collaborators provided material. It is difficult, therefore, to identify precisely the contributions made by individuals and we apologise for any oversight in the following list of principal contributors.

François Cupelin, Service of Ecotoxicology (ECOTOX), Geneva, Switzerland (Chapter 7)

Peter G. Flachsbart, University of Hawaii at Manoa, Honolulu, Hawaii (Chapter 4)

David T. Mage, United States Environmental Protection Agency, North Carolina, USA (Chapter 6)

Isabelle Romieu, World Health Organization/Pan American Health Organization, Mexico City, Mexico, now at the Center for Disease Control, Atlanta, Georgia, USA (Chapter 2)

Ragnar Rylander, University of Gothenburg, Sweden (Chapter 3)

Dietrich Schwela, World Health Organization, Geneva, Switzerland (Chapters 1 and 8)

Aswani K. Singh, Service of Ecotoxicology (ECOTOX), Geneva, Switzerland, now at APEX Consulting Services, Thônex (Chapter 1)

Michael P. Walsh, Consultant, Arlington, Virginia, USA (Chapters 1, 5 and 6)

Olivier Zali, Service of Ecotoxicology (ECOTOX), Geneva, Switzerland, now at the Département de l'action sociale et de la santé, République et Canton de Genève, Geneva, Switzerland (Chapters 1 and 7)

Thanks are due to Philipp Schwela for secretarial services, for administrative assistance in the preparation of the original manuscript and for assistance with the preparation of the final manuscript. We are also grateful to Alan Steel for preparation of illustrations, to Leonard Chapman for assistance with production and layout and to Stephanie Dagg for compiling the index. Thanks are also due to Deborah Chapman for language editing and for acting as series editor and production manager in collaboration with the publisher. As the editor of the WHO co-sponsored series of books dealing with various aspects of environmental health management, she was also responsible for ensuring compatibility with other books in the series.

Chapter 1*

MOTOR VEHICLES AND AIR POLLUTION

In their efforts to achieve sustainable development, governments world-wide are facing a growing problem of the health effects induced by air pollutant concentrations arising from motor vehicle emissions.

This book examines three aspects of the problem of air pollution caused by motor vehicles:

- The effects of traffic emissions and noise on health (Chapters 2 and 3).
- Human exposure to the emitted pollutants (Chapter 4).
- Control actions for limiting the emission of the pollutants and the effectiveness of such control actions (Chapter 5).

Case studies of the different ways in which the problem of motor vehicle air pollution has been addressed in developing and developed countries are presented in Chapters 6 and 7.

Only those issues relating directly to motor vehicles are considered here. Those that are indirectly linked are not discussed, most notably the contribution of motor vehicles to global carbon dioxide (CO_2) production, the related greenhouse effect, the role of infrastructure (road network design and construction) and land-use planning.

The effect of noise from motor vehicles on health is considered in Chapter 3 but control actions to limit this form of pollution are not. Traffic accidents, which represent a serious public health problem, are also beyond the scope of this report. It should be noted, however, that limiting the number of vehicles on the road by improving mass transport systems will improve road safety and reduce the rate of traffic accidents, and will also help to reduce air pollution.

The overall aim of this book is to provide a basis for informed decision making on how to control automotive air pollution in both developing and developed countries. The influence of the geographical and socio-economic situation is emphasised throughout and descriptions of the situation in a variety of cities of widely different character are provided.

* *This chapter was prepared by Dietrich Schwela and Olivier Zali*

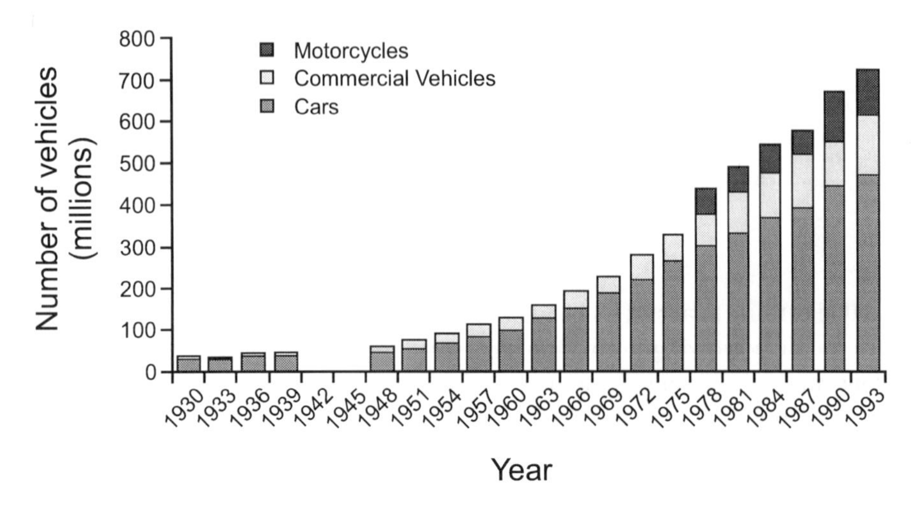

Figure 1.1 Growth in world total numbers of cars, motorcyles and commercial vehicles between 1930 and 1993 (Data from the Motor Vehicle Manufacturers' Association, 1994)

1.1 Current situation

Across the entire globe, motor vehicle traffic has increased tremendously. In 1950, there were about 53 million cars on the world's roads; 44 years later, the global private car fleet had grown to 460 million, a nine-fold increase. On average, the fleet has grown by 9.5 million units per year over this period. Simultaneously, as illustrated in Figure 1.1, the lorry (truck) and bus fleet has been growing by about 3.6 million vehicles per year (Motor Vehicle Manufacturers' Association, 1994). While this growth rate has slowed considerably in the industrialised countries, population growth and increased urbanisation and industrialisation have accelerated the use of motor vehicles elsewhere. If the two-wheeled vehicles around the world are included (estimated at 120 million and growing at about 4 million vehicles per year over the last decade), the global motor fleet in 1994 was about 715 million.

Looking at the present global vehicle numbers (Figure 1.2 and Table 1.1), it is clear that there are wide disparities between regions of the world and their types of vehicle. Figure 1.2A shows the distribution of private car registrations in various regions of the world. Eighty-six per cent are concentrated in developed countries (Europe, North America and Japan), while Asia (excluding Japan) and Africa, with more than half of the world's population, have about 7 per cent. The proportions are similar when commercial vehicles are included (i.e. buses, coaches, trucks and vans) — developed countries

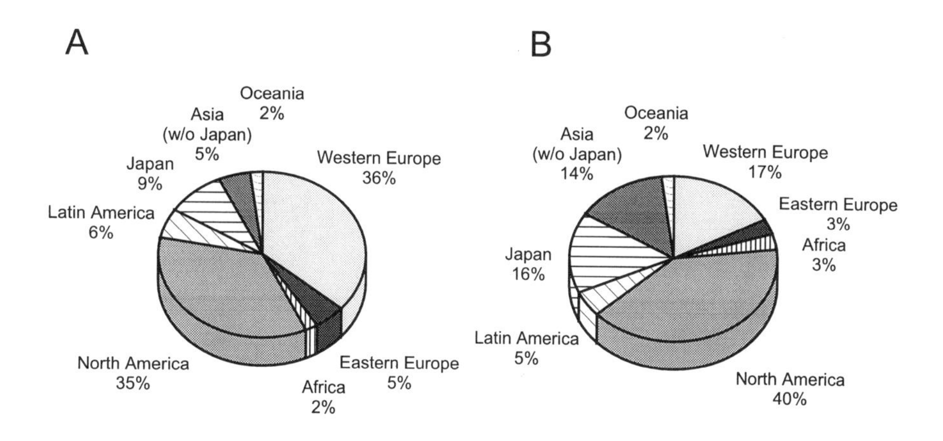

Figure 1.2 Global distribution of vehicles, 1994: **A.** private cars; **B.** all motor vehicles (Data from International Road Federation, 1995)

account for 76 per cent, while for Asia and Africa the level rises to 17 per cent. Most of the world's population still live and work largely without motorised transport.

Table 1.1 shows the enormous variation in motorisation from one region of the world to another, ranging from less than 14 private cars (for Asia excluding Japan) to more than 550 private cars (for North America) per 1,000 inhabitants. In certain Asian or African countries, the level of motorisation is even less than one per 1,000 inhabitants. Consequently, if developing countries reach the degree of motorisation found in Europe or North America (sometime in the future), the problems related to the enhanced concentrations of air pollution in urban areas will be greatly exacerbated. Such problems may be compounded by the fact that, in developing countries, petrol (gasoline) often has a high sulphur content and engines are not always properly tuned. Moreover, the proportion of diesel-powered trucks in the vehicle fleet, and of two-wheelers powered by two-stroke engines, strongly influences air quality. Taiwan had 562 two-wheelers per 1,000 persons in 1994, three and a half times more than other types of automobiles. Similar high proportions of two-wheelers exist in Malaysia, Thailand and Indonesia, and very high levels of growth are currently occurring in China, India and Vietnam. Furthermore, in cities that are subject for long periods of the year to continental high-pressure systems, such as Beijing, Delhi or Mexico, smog is becoming a serious issue.

Table 1.1 Distribution of private cars, commercial vehicles and motorised two-wheelers by major geographical region, 1994

	Human population	Private cars	Commercial vehicles	Two wheelers	Total vehicles
		Total numbers (millions)			
Western Europe	448.9	161.3	26.6	19.5	207.5
Eastern Europe	170.4	24.3	3.9	4.0	32.1
Africa	362.5	8.1	3.8	0.9	12.8
North America	286.5	159.8	56.2	4.2	220.2
Latin America	349.8	28.5	6.3	0.7	35.5
Japan	125.0	42.7	22.4	8.9	74.0
Asia (w/o Japan)	1,577.6	22.8	19.4	37.8	80.0
Oceania	21.2	9.7	2.4	0.3	12.4
World	3,341.8	457.1	141.1	76.3	674.6
		Number per thousand population			
Western Europe		359	59	44	462
Eastern Europe		142	23	23	188
Africa		22	11	2	35
North America		558	196	15	769
Latin America		81	18	2	101
Japan		341	180	71	592
Asia (w/o Japan)		14	12	24	51
Oceania		456	115	16	587
World		137	42	23	202

Data do not include China and Russia

Source: International Road Federation, 1995

1.2 Emissions of atmospheric pollutants

Comprehensive data on air pollution emissions from transportation and other activities are not available for all countries. However, the data published for the Organisation for Economic Cooperation and Development (OECD) countries (OECD, 1993), as well as for several other European countries (Poland, Czech Republic, Hungary), illustrate the major trends (Table 1.2). Transport accounts for about 4 per cent of sulphur dioxide (SO_2) emissions and its contribution is tending to decrease; the situation appears to be under control, at least for developed countries. By contrast, transportation contributes a higher proportion of the emissions of other atmospheric pollutants.

For nitrogen oxides (NO_x), the proportion of emissions due to transport is more than 50 per cent. This level hardly changed between 1980 and 1990, although the overall emissions declined by 26 per cent over the same period. It is hoped that a further decline will occur in future years as the number of vehicles equipped with catalytic converters increases in Western Europe and Japan. Quantities of particulate matter emissions have remained stable, but

Table 1.2 Pollutant emissions from vehicles and other sources in OECD countries 1980 and 1990

Pollutant	Emission source	Total emissions (10^3 t)	
		1980	1990
SO_2	Vehicles	2,144	1,664
	Fixed sources	60,075	38,372
NO_x	Vehicles	21,613	15,845
	Fixed sources	19,629	14,538
Particles	Vehicles	1,967	1,998
	Fixed sources	16,038	12,512
CO	Vehicles	122,440	72,824
	Fixed sources	40,726	31,260
VOC	Vehicles	14,309	7,947
	Fixed Sources	19,871	17,890

Source: OECD, 1993

the portion attributable to motor transportation has increased slightly, from 11 per cent to 14 per cent due to the strong growth in diesel-powered road transport and to successes in reducing particulate emissions from fixed sources. Between 1980 and 1990 the proportion of carbon monoxide (CO) emitted by motor vehicles decreased from 75 per cent to 70 per cent. More significantly, the total CO emissions declined by 36 per cent during the same period. Finally, considerable progress has been achieved in controlling volatile organic compounds (VOCs); emissions attributable to motorised traffic fell from 14×10^6 t in 1980 (42 per cent) to 8×10^6 t in 1990 (30 per cent). Despite these improvements, CO concentrations are still too high. In urban areas with the highest pollution levels, the motor vehicle contribution tends to increase. Usually, over 90 per cent of CO in city centres comes from vehicles and it is common to find 50–60 per cent of the hydrocarbons (HC) and NO_x also originating from vehicles. Chapter 4 discusses the various pollutants from motor vehicles in greater detail, particularly with respect to the number of persons exposed to atmospheric pollutants.

1.3 Future trends

The number of motor vehicles world-wide is growing by about 5 per cent per year, far faster than the global human population which has been increasing by about 2 per cent per year. Analysis of the trends in global motor vehicle

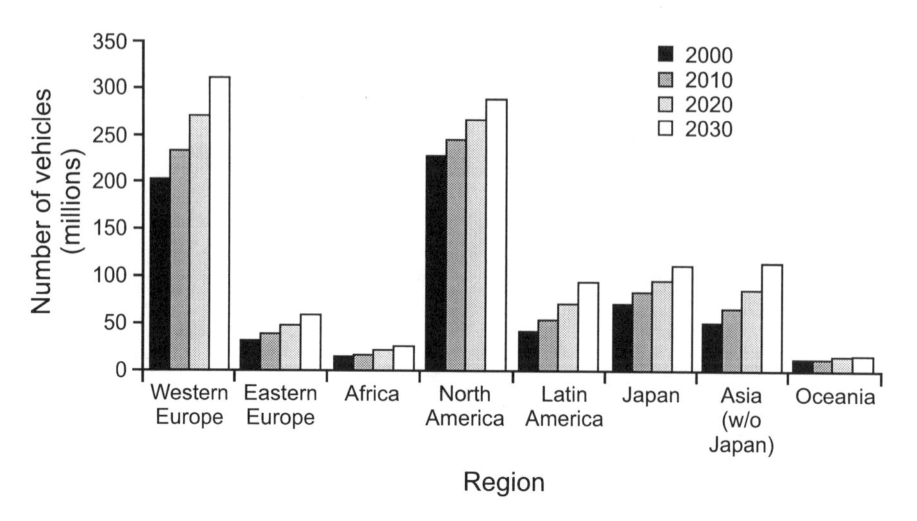

Figure 1.3 Estimated growth in the total numbers of vehicles by world region (Data provided by Mike Walsh, Arlington, VA, 1995)

registrations has revealed that the global fleet has been growing by approximately 16 million units per year, not including two-wheeled vehicles.

World-wide registrations have been growing by about 1.8 cars per 1,000 persons or 2.3 vehicles (cars plus trucks plus buses) per 1,000 persons. If this trend were to continue until 2010, there would be 215 motor vehicles per 1,000 persons (excluding motorcycles) compared with 178 in 1994. Countries with large populations and low degrees of motorisation, such as China, are not included in these statistics and therefore world averages would probably be lower by 20–30 per cent if these countries were taken into account.

Projections of the future vehicle population have been made, taking into consideration population growth and economic development, which are the two most important factors influencing growth in vehicle numbers. Estimates of total vehicle numbers (cars, trucks, buses and motorcycles) up to the year 2030 are summarised in Figure 1.3. Some decline in the growth rate is noticeable because these projections take into account the increasing saturation of the motor vehicle market in developed regions, as well as the probable effects of policy interventions by governments to restrain future growth in the numbers of motor vehicles, particularly in highly industrialised, urbanised and polluted regions. In spite of these factors, the number of vehicles per 1,000 persons is expected to continue to increase in all regions of the world, although developed countries will continue to be responsible for the greatest proportion of vehicle numbers. The increase in the number of vehicles in Africa, and in a large part of Asia, will remain small. However, atmospheric pollution caused by motor vehicles is already a serious problem in the large cities of these regions.

The global vehicle fleet has tended to be dominated by the highly industrialised areas of North America, Western Europe and Japan. This pattern is gradually changing, not because the numbers of vehicles in these areas have stopped increasing, but because the rate of increase is accelerating in other areas. By 2030, based on current trends, the regions of Asia (excluding Japan), Eastern Europe, Africa and Latin America will represent 28 per cent of motor vehicle registrations compared with 20 per cent at present, although per capita growth rates will remain low.

In conclusion, important decisions need to be taken urgently concerning motor vehicle induced atmospheric pollution in order to minimise the future costs of effects on health and on the environment. These decisions relate, in particular, to land-use planning, public transportation systems and fuel composition. In view of the importance of the automobile, choices made in these areas will influence development strongly over the next 20–50 years. It is intended that the information contained in this book will support the decisions that will lead to vital initiatives in pollution prevention from motor vehicle emissions.

1.4 References

International Road Federation 1995 *World Road Statistics 1990-1994.* International Road Federation, Geneva and Washington D.C.

Motor Vehicle Manufacturers' Association 1994 *1993 World Motor Vehicles Data.* Motor Vehicle Manufacturers' Association of the United States, Inc., Detroit, Michigan.

OECD 1993 *OECD Environmental Data Compendium 1993.* Organisation for Economic Cooperation and Development, Paris.

Chapter 2*

EPIDEMIOLOGICAL STUDIES OF HEALTH EFFECTS ARISING FROM MOTOR VEHICLE AIR POLLUTION

Several major classes of air pollutants have the potential to affect the health of populations. These pollutants result either from primary emissions or atmospheric transformation. Motor vehicles are the major source of a number of these pollutants, in particular, carbon monoxide, nitrogen oxides, unburned hydrocarbons and lead and, in smaller proportions, suspended particulate matter, sulphur dioxide and volatile organic compounds.They also contribute via atmospheric transformation to increases in ozone and other photochemical oxidants (HEI, 1988). With growing urbanisation and increases in vehicle density, urban air pollution has become a crucial problem, compounded by the great expense of pollution control, and it is now urgent to undertake risk assessments in order to evaluate and prioritise control strategies.

There are an extremely large number of substances in the exhaust of motor vehicles and which are capable of causing health effects. Such substances fall into three main categories:

- Substances mainly affecting the airways, i.e. NO_2, ozone (O_3), photochemical oxidants, SO_2 and suspended particulate matter (SPM),
- Substances producing toxic systemic effects (CO, Pb).
- Substances with potential carcinogenic effects (benzene, polycyclic aromatic hydrocarbons (PAHs), and aldehydes).

This chapter reviews the epidemiological evidence of the health effects of pollutants that have a direct or indirect source relationship to automotive emissions. Individuals exposed to a high concentration of air pollution resulting from road traffic are, at the same time, exposed to high noise levels; the health effects of the latter are discussed in Chapter 3.

2.1 Factors governing the toxic effects of pollutants

Motor vehicle exhaust is a complex mixture, the composition of which depends on the fuel used as well as on the type and operating condition of the

* *This chapter was prepared by Isabelle Romieu*

engine and whether it uses any emission control device. Pollutants and their metabolites can cause adverse health effects by interacting with, and impairing, molecules crucial to the biochemical or physiological processes of the human body. Three factors govern the risk of toxic injury from these substances: their chemical and physical properties, the dose of the material that reaches critical tissue sites, and the responsiveness of these sites to the substance. The physical form and properties (e.g. solubility) of airborne contaminants influences their distribution in the atmosphere and in biological tissues, and therefore the dose delivered to the target site. This dose is very difficult to determine in epidemiological studies and therefore surrogate measurements are used ranging from atmospheric concentration to dose determination in blood or more accessible body tissues (for example, hair). For some pollutants, mathematical models of the relationship between exposure and dose can also be used to develop surrogate measurements.

The interaction of pollutants with biological molecules (or receptors) triggers the mechanism of toxic response that may act by direct stimulation, or by a cascade of molecular and cellular events that ultimately damage tissue (HEI, 1988). The different pathways of pollutant sources, from exposure through inhalation, to toxic effects are shown in Figure 2.1.

Pollutant effects may also vary across population groups; in particular, the young and the elderly may be especially susceptible to deleterious effects; persons with asthma or other pre-existing respiratory or cardiac diseases may experience aggravated symptoms upon exposure (HEI, 1988).

2.2 Nitrogen dioxide

Nitrogen dioxide is an irritating gas and its toxicity is generally attributed to its oxidative capabilities. It penetrates the lung periphery and is primarily deposited in the centriacinar region. It is also absorbed into the mucosa of the respiratory tract.

Upon inhalation 80–90 per cent of NO_2 can be absorbed, although this proportion varies according to nasal or oral breathing. The maximal dose to the lung tissue occurs at the junction of the conducting airway and the gas exchange region. Because NO_2 is not very soluble in aqueous surfaces, the upper airways retain only small amounts of inhaled nitrogen oxides. Nitric and nitrous acids or their salts can be observed in the blood and urine after exposure to NO_2 (WHO, 1987a). Nitrogen dioxide is a widespread contaminant of indoor as well as outdoor air, and indoor levels can exceed those found outdoors. Indoor levels of NO_2 are determined by the infiltration of NO_2 from outdoor air and by the presence and strength of indoor sources, such as gas cooking stoves and kerosene space heaters, and by air exchange (Bascom *et al.,* 1996).

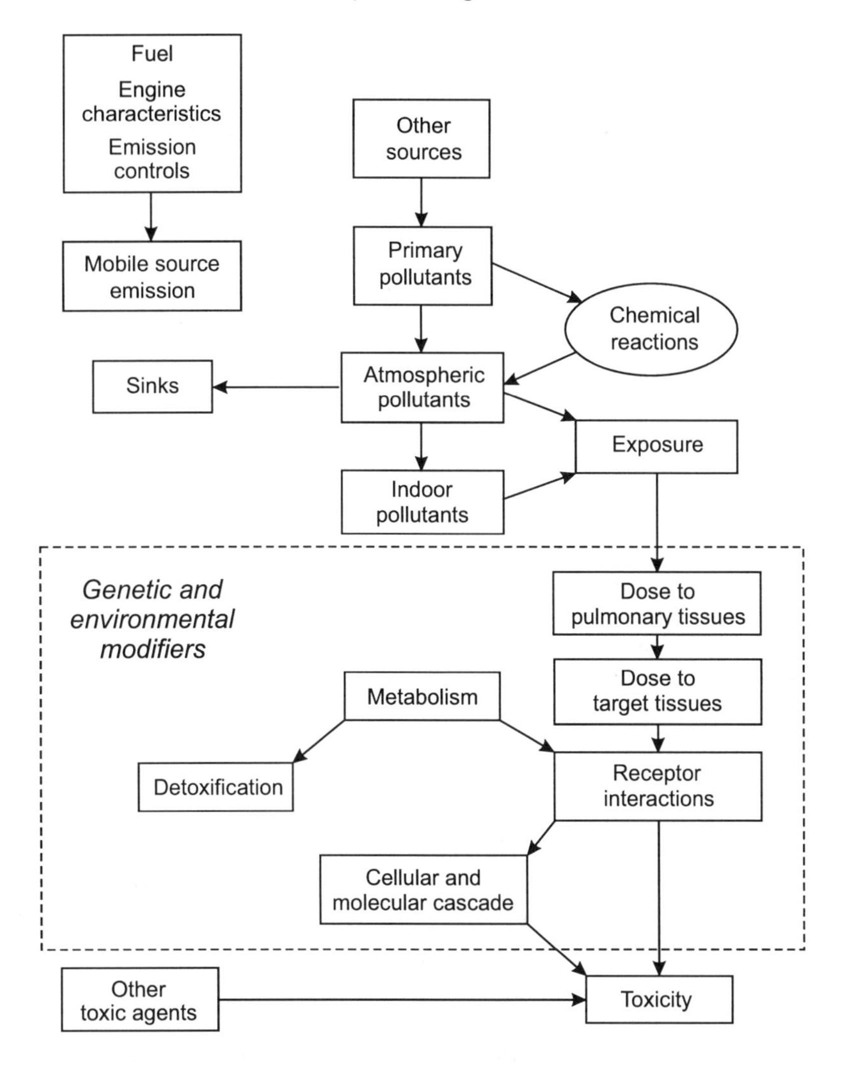

Figure 2.1 Pathways from motor vehicle pollutant sources to toxic effects in humans by exposure through inhalation (After HEI, 1988)

In certain occupations, workers are intermittently exposed to high concentrations of oxides of nitrogen, particularly nitrogen oxide NO and NO_2. The spectrum of pathological effects in the lung resulting from occupational exposure to nitrogen oxides range from mild inflammatory response in the mucosa of the tracheobronchial tree at low concentrations, to bronchitis, to bronchopneumonia, and to acute pulmonary oedema at high concentrations (WHO, 1977).

Most of the epidemiological research studies at the community level have focused on the acute effects of short-term exposure to high levels of NO_2, and there are few data on long-term effects of low level or repeated exposure at peak levels. Morrow (1984) has shown that NO_2 can be toxic in certain biological systems, and acute exposure to NO_2 has been reported to affect both the cellular and humoral immune systems. Devlin *et al.* (1992) reported an impaired phagocytic activity in human alveolar macrophages after exposure to 3,760 μg m^{-3} (2.0 parts per million (ppm)) for 4 hours, with intermittent exercise. Damji and Richters (1989) have reported a reduction of T-lymphocyte subpopulations following acute exposure to NO_2, and this may reflect a functional impairment of the immune response.

There have been numerous controlled studies of the effect of nitrogen dioxide on the lung functions of healthy individuals, asthmatics and subjects with chronic bronchitis (US EPA, 1982a). Short exposure (10–15 minutes) to concentrations of NO_2 exceeding 1,300 μg m^{-3} (0.7 ppm) caused functional changes in healthy subjects, particularly an increased airway resistance. Recent controlled studies show conflicting results concerning respiratory effects in asthmatics and healthy individuals at NO_2 concentrations in the range of 190–7,250 μg m^{-3} (0.1–4.0 ppm). The lowest observed level to affect lung function consistently was a 30-minute exposure, with intermittent exercise, to a NO_2 concentration of 380–560 μg m^{-3} (0.2–0.3 ppm). Asthmatics appear to be more responsive to NO_2, and their lung functions may be affected by a concentration of 560–940 μg m^{-3} (0.3–0.5 ppm) with an enhanced reactivity to pharmacological bronchoconstrictor agents (Mohsenin, 1987a; WHO, 1995a). However, several studies have failed to demonstrate an adverse effect of NO_2 on the respiratory health of asthmatics (Morrow and Utell, 1989; Roger *et al.*, 1990). Inconsistencies in the results could be due to the lack of comparability in experimental studies and a difference in susceptibility among asthmatics.

Few community epidemiological studies of outdoor NO_x exposure have demonstrated an association between ambient air levels of NO_x compounds and measurable health effects. However, methodological problems, such as the presence of a mixture of pollutants and a lack of control due to parental smoking or indoor sources of NO_2 (Speizer *et al.*, 1980), in all of the studies preclude acceptance of any of the results as clear evidence for an increase in acute respiratory illness due to NO_2 exposures. These studies have been extensively reviewed by the United States Environmental Protection Agency (US EPA) (US EPA, 1982a). To summarise the results of studies reported up until 1990, a meta-analysis of 11 epidemiological studies showed an increase in respiratory illness in children of less than 12 years of age, associated with

Table 2.1 Potential human health effects of NO_2

Health effect	Mechanism
Increased intensity of respiratory infections	Reduced efficacy of lung defences
Increased severity of respiratory infections	Reduced efficacy of lung defences
Respiratory symptoms	Airways injury
Reduced lung function	Airways and alveolar (?) injury
Worsening of the clinical status of persons with asthma, chronic obstructive pulmonary disease or other chronic respiratory conditions	Airways injury

Source: Samet and Utell, 1990

long-term exposure to high concentrations of NO_2 (gas stoves), compared with children exposed to low concentrations of this pollutant. A difference in exposure of 28.5 μg m^{-3} (0.015 ppm) NO_2 (2-week average) resulted in an increase of about 20 per cent in the odds ratio of respiratory illness (odds ratio of 1.2 with a 95 per cent confidence interval between 1.1 and 1.3). In these studies, the weekly NO_2 average ranged from 15 to 128 μg m^{-3} (8–65 ppb) or possibly higher. There was, however, insufficient epidemiological evidence to draw any conclusions regarding the short- or long-term effects of NO_2 on pulmonary function (Hasselblad *et al.*, 1992).

More recent studies (Dijkstra *et al.*, 1990; Neas *et al.*, 1991; Samet *et al.*, 1993) have concentrated on indoor air pollution through NO_2. These studies could not establish a consistent trend in incidence or duration of illness due to NO_2 exposure. In addition no consistent effect of NO_2 on pulmonary function could be confirmed, and no association was observed between symptoms such as chronic cough, persistent wheeze and shortness of breath, with NO_2 concentrations. In one study (Braun-Fahrlander *et al.*, 1992) the duration of episodes with symptoms was associated with outdoor, but not indoor, NO_2 concentrations. The inconsistency in the results of epidemiological studies has been commented on by Samet and Utell (1990) who pointed out the potential for misclassification, confounding and lack of statistical power among the investigations. Also, some factors such as exposure pattern, age, nutritional status, and interaction with other pollutants or allergen, may partly explain the variability observed in response to NO_2 exposure.

In summary, in spite of decades of laboratory, clinical and epidemiological research, the human health effects of NO_2 exposure have not been fully characterised (Table 2.1). The toxicological evidence has indicated hypotheses to be tested in human populations but limitations in the clinical and epidemiological studies have precluded such definitive testing of these hypotheses. There is a need to characterise factors that may modulate response to NO_2 exposure (Samet and Utell, 1990).

The following guidelines have been proposed by WHO (WHO, 1995a): a 1-hour guideline of 200 μg m^{-3} (0.11 ppm) and an annual guideline of 40 μg m^{-3} (0.021 ppm).

2.3 Ozone and other photochemical oxidants

The primary target organ for O_3 is the lung. Ozone exposure produces cellular and structural changes, the overall effect of which is a decrease in the ability of the lung to perform normal functions. Ciliated and Type 1 cells are the most sensitive to O_3 exposure (ciliated cells clear the airway of inhaled foreign material). Proliferation of non-ciliated bronchiolar and Type 2 alveolar cells occurs as a result of damage and death of ciliated and Type 1 cells. Ozone exposure causes major lesions in the centriacinar area of the lung which includes the end of the terminal bronchioles and the first few generations of either respiratory bronchioles or alveolar ducts (depending on the species of animal) (Lippmann, 1989a).

The observed health effects of photochemical oxidant exposure cannot be attributed uniquely to oxidants because photochemical smog typically consists of O_3, NO_2, acid sulphate and other reactive agents. These pollutants may have additive or synergistic effects on human health, but O_3 appears to be the most biologically active (WHO, 1978, 1987b).

Experimental studies in animals and humans have shown that O_3 increases airway permeability and particle clearance, causes airway inflammation and a decrease in bactericidal capacity, and causes structural alterations in the lung. The acute morphologic response to O_3 involves epithelial cell injury along the entire respiratory tract, resulting in cell loss and replacement. In the lower airway, the proximal regions are most affected. In the alveolus, Type I cells are highly sensitive to O_3 while Type II cells are resistant and appear to serve as a stem cell for Type I cell replacement (Lippmann, 1989b). Studies that have examined the composition of the bronchial alveolar lavage (BAL) have reported an increase in numbers of neutrophils (Schelegle *et al.*, 1991; Koren *et al.*, 1991) and in cytokine chemotactic for neutrophils (LDH and IL-8) (Aris *et al.*, 1993). Ozone-induced increases in constituents of BAL fluid such as lactate dehydrogenase, prostaglandine E2, interleukin-6, fibronectin and albumin, have been observed in humans (Devlin *et al.,* 1991). The increase in the content of neutrophils in nasal lavage fluid has also been observed in humans after exposure to O_3 (Graham and Koren, 1990;

Smith *et al.*, 1993). It is not yet known whether repetitive inflammation has long-term consequences.

Ozone, like NO_2, can induce increased non-specific airway sensitivity to inhalation challenge testing with bronchoconstrictive agents (HEI, 1988). In several controlled human studies, reductions in lung function have been observed among voluntary subjects exposed to O_3 concentrations ranging from 0.12 ppm to 0.40 ppm (McDonnell *et al.*, 1983; Avol *et al.*, 1984; Linn *et al.*, 1986). Recent research has shown that effects can be produced by exposures as short as 5 minutes, and that various effects become progressively worse as exposures at a given concentration are extended in time up to 6.6 hours. Exposure to 160 $\mu g\ m^{-3}$ (0.08 ppm) for 6.6 hours in a group of healthy exercising adults led to a decrease in lung functions of more than 10 per cent in the most sensitive individuals. However, repeated exposures to a given concentration (6.6 hours to 0.08, 0.10 and 0.12 ppm) on several consecutive days resulted in attenuation of functional changes but persistence of airway hyper-responsiveness. This suggests an ongoing action of O_3 on the lung (Folinsbee *et al.*, 1994). Although smokers and subjects with pre-existing pulmonary disease do not appear to be more sensitive than others to O_3, within the apparently normal population there is a range of responsiveness to O_3 that is reproducible (WHO, 1987b).

Most studies of the health effects of O_3 have focused on short-term (1–2 hour) exposure and have indicated a number of acute effects of O_3 and other photochemical oxidants (Table 2.2). This literature has been reviewed by the US EPA (US EPA, 1986b).

Recently, some studies have linked acute daily mortality with O_3 exposure in Los Angeles County and New York City (Kinney and Ozkaynat, 1991, 1992), but data from a study conducted in Mexico City do not confirm these results (Borja-Aburto *et al.*, 1995; Loomis *et al.*, 1996). Studies of hospital admissions in relation to O_3 exposure reported an increase in hospital admission rates for respiratory diseases (Bates and Sizto, 1983, 1987) and asthma attacks (Whittemore and Korn, 1980; White and Etzel, 1991; White *et al.*, 1994; Romieu *et al.*, 1995). Summer haze pollutants (including O_3) have also been related to hospitalisation and emergency room visits for respiratory diseases (Cody *et al.*, 1992; Thurston *et al.*, 1992, 1994).

However, due to methodological limitations, these studies cannot fully demonstrate an association between O_3 exposure and respiratory illness. It is also difficult to determine the effect of a single pollutant because subjects are normally exposed to a complex air pollution mixture.

Different studies have investigated symptoms related to O_3 exposure. In a clinical study, Avol *et al.* (1987) studied the occurrence of symptoms in

Table 2.2 Human responses to a single O_3 exposure

Response	Subjects	Exposure conditions
5–10 % mean decrement in FEV	Healthy young men	10 ppb with intermittent heavy exercise for 2h — O_3 in purified air; 100 ppb with moderate exercise for 6.6 h — O_3 in purified air; 100 ppb with very heavy exercise for 0.5 h — O_3 in ambient air
	Healthy children	100 ppb normal summer camp programme — O_3 in ambient air
Increased cough	Healthy young men	120 ppb with intermittent heavy exercise for 2h — O_3 in purified air
	Healthy young men	80 ppb moderate exercise for 6.6 h — O_3 in purified air
	Healthy young men and women	120–130 ppb heavy exercise for 16–28 min — O_3 in purified air
Reduced athletic performance	Healthy young men	180 ppb with exercise at V_E of 54 litres/min for 30 min, 120 litres/min for 30 min — O_3 in purified air
	Healthy young men and women	120–130 ppb with exercise at V_E of 30–120 litres/min for 16–28 min — O_3 in purified air
Increased airway reactivity	Healthy young men	80 ppb with moderate exercise for 6.6h — O_3 in purified air
	Healthy young adult men with allergic rhinitis	180 ppb with heavy exercise for 2h — O_3 in purified air
Increased airways permeability	Healthy young men	400 ppb with intermittent heavy exercise for 2h — O_3 in purified air
Increased airway inflammation	Healthy young men	80 ppb with moderate exercise for 6.6h — O_3 in purified air
Accelerated tracheo-bronchical particle clearance	Healthy young men	200 ppb with intermittent light exercise for 2h — O_3 in purified air

1 ppb (parts per billion) = 2 μg m^{-3}
Source: Lippmann, 1989b

FEV Forced Expiratory Volume in 1 second

subjects exposed to O_3 concentrations of 0–640 μg m^{-3} (0–0.32 ppm). Symptoms were classified as upper respiratory (nasal congestion or discharge and throat irritation), lower respiratory (substernal irritation, cough, sputum production, dyspnea, wheeze and chest tightness) and non-

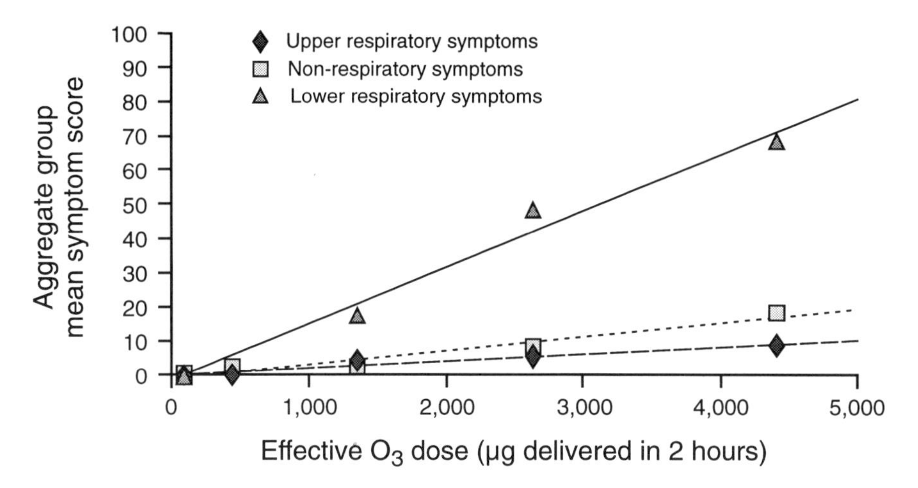

Figure 2.2 Effects of ozone on respiratory systems (After Kleinman *et al.*, 1989a)

respiratory (headache, fatigue and eye irritation). Scores were calculated, based on the intensity of the symptoms presented. The results showed a dose-response relationship between the effective dose of O_3 (O_3 concentration × time × ventilation rate) and symptom scores (Figure 2.2).

Imai *et al.* (1985) reported significant increases in the symptoms of adults during periods of increased ambient O_3 exposure in Japan. Ostro (1989) analysed data collected for 6 years during the National Health Interview Survey, 1976–81, and showed that ambient O_3 levels were associated with restricted days of activity due to respiratory illness in the working population. According to this analysis, the change in the number of days of minor or restricted activity of an individual on a given day in a given population, is proportional to the product of the daily extreme value of the hourly O_3 concentration on a given day and the size of the exposed population; the proportionality factor is 0.077 (Kleinman *et al.*, 1989a).

Although O_3 has been shown to alter macrophage function, which could lead to an increased susceptibility to respiratory infection, there is little data on the relation of exposure to O_3 to acute respiratory infection in humans. In a study conducted in Mexico among preschool children, investigators reported a 14 per cent increase (odds ratio of 1.14 with a 95 per cent confidence interval between 1.10 and 1.18) in school absenteeism for respiratory infection related to an O_3 ambient concentration (1 h maximum) > 240 μg m^{-3} (0.12 ppm) on the preceding day (lag of 1 day). When O_3 concentrations (1 hour maximum) were high (> 240 μg m^{-3}) on two consecutive days, a 22 per cent increase was observed (odds ratio of 1.22 with a 95 per cent confidence interval between

1.16 and 1.28). An interactive effect was observed between low temperature and high O_3 exposure (Romieu *et al.*, 1992).

The inhalation of O_3 causes concentration-dependent decreases in average lung volumes and flow rates during expiration; the mean value of the decrease increases with increasing depth of breathing (Lippmann, 1989a). Decreases in the lung functions of healthy children and young adults have been reported at hourly average O_3 concentrations in the range 160–300 µg m^{-3} (0.08 –0.150 ppm) (Table 2.3).

Based on estimates from Spektor *et al.* (1988), moderate physical activity for a range of exposure between 38 and 226 µg m^{-3} (0.019 to 0.113 ppm) for 1 hour could lead to a decrease of 0.5 ml per µg m^{-3} for FVC (Forced Vital Capacity) and 0.7 ml per µg m^{-3} for FEV_1 (Forced Expiratory Volume in 1 second); this would result in a decrease of 180 ml for FVC and 250 ml for FEV_1 for a concentration of 400 µg m^{-3} (0.2 ppm) O_3 (1-hour average). From these data, the average decreases in FVC, FEV_1 and PEFR (Peak Expiratory Flow Rate) of 4.9 per cent, 7.7 per cent and 17 per cent respectively were predicted for the current US EPA standard of 240 µg m^{-3} (0.12 ppm). Further studies conducted among children have shown similar decreases in lung function and have reported the persistence of a measurable functional deficit into the following day, even for peak concentrations below or equal to 300 µg m^{-3} on the day before (0.15 ppm) (Lippman, 1993).

In a study conducted in Mexico City among school children, Castillejos *et al.* (1992) reported acute and subacute effects of O_3 on lung functions. However, the decreases in function were smaller than was to be expected from the regression slope of Spektor *et al.* (1988) (0.8 per cent for FVC and 0.8 per cent for FEV_1 at a maximum O_3 concentration of 240 µg m^{-3} (0.12 ppm) 24 h prior to testing). Mean O_3 exposures of 48 h and 168 h (7 days) were more significant in predicting FEV_1 and FEF_{25-75} (Forced Expiratory Flow from 25 per cent to 75 per cent of the Forced Vital Capacity). These authors suggested that children chronically exposed to O_3 may present a phenomenon of "tolerance". This finding supports the observation that repetitive exposures tend to produce less of a response (Folinsbee *et al.*, 1994). The potential adverse effect of such "tolerance" is not known.

Recently, Avol *et al.* (1998) studied asthmatic, wheezy and healthy children in Southern California during periods of varying O_3 levels in spring and late summer. The study found that asthmatic children had the most trouble breathing, the most wheezing and the most use of inhalers during high O_3 days in the spring. Wheezy children had the most trouble breathing and wheezed more during low O_3 days in summer.

Table 2.3 Changes in forced vital capacity and expiratory flow and volume in adults and children after exercise at different O_3 concentrations during field and chamber studies

Study	Number of subjects and gender	Age range or mean value (years ± s.d.)	Activity level (min. vent. in litres ± s.d.)	Exposure/ exercise time (mins)	Mean rate of functional change: O_3 conc. (ppb)	FVC (ml/ppb)	FEV_1 (ml/ppb)	PEFR (ml s^{-1}/ppb)	FE_{25-27} (ml s^{-1}/ppb)
Adults									
Folinsbee *et al.* (1988)	10 M	18–33	Moderate (40)	395/300	120 [1]	–3.8	–4.5		–5.0
Gibbons and Adams (1984)	10 F	22.9 ± 2.5	High (55)	60/60	150 [1]	–1.1	–1.0		–0.6
Avol *et al.* (1984)	42 M,8 F	26.4 ± 6.9	High (57)	60/60	153 [2]	–1.2	–1.3		
					160 [1]	–1.5	–1.5		
McDonnell *et al.* (1983)	22 M	22.3 ± 3.1	High (65)	120/60	120 [1]	–1.4	–1.3		–2.9
	20 M	23.3 ± 3.2	High (65)	120/60	180 [1]	–1.8	–1.6		–3.0
Kulle *et al.* (1985)	20 M	25.3 ± 4.1	High (68)	120/60	150 [1]	–0.5	–0.2		–2.1
Linn *et al.* (1986)	24 M	18–33	High (68)	120/60	160 [1]	–0.7	–0.6	–1.1	–1.1
Spektor *et al.* (1988)	20 M, 10 F	22–44	Varied (78.6 ± 34.8)	29.3 ± 9.2	21–124 [2]	–2.1	–1.4	–9.2	–6.0
	7 M, 3 F	22–40	(64.6 ± 10.0)	26.7 ± 8.7	21–124 [2]	–2.9	–3.0	–13.7	–9.7
Children									
Lioy *et al.* (1985)	17 M, 22 F	7–13	Low	150–550	20–145 [2]	–0.1	–0.3	–3.0	–0.6
Kinney *et al.* (1988)	94 M, 60 F	10–12	Low	1,440	7–78 [2]	–0.9	–1.0 [3]		–1.9
Lippmann *et al.* (1983)	34 M, 24 F	8–13	Moderate	150–550	46–110 [2]	–1.1	–0.8		
Spektor *et al.* (1988)	53 M, 38 F	7–13	Moderate	150–550	19–113 [2]	–1.0	–1.4	–6.8	–2.5
Avol *et al.* (1988)	33 M, 33 F	8–11	Moderate (22)	60 (60)	113 [2]	–0.3	–0.3	–0.4	
Avol *et al.* (1985)	46 M, 13 F	12–15	High (32)	60 (60)	150 [1]	–0.7	–0.8	–1.6	–0.7
McDonnell *et al.* (1985)	23 M	8–11	Very high (39)	150 (60)	120 [1]	–0.3	–0.5	–1.8	–0.6

s.d. Standard deviation
FEV_1 Forced expiratory volume in one second
FVC Forced vital capacity
PEFR Peak expiratory flow rate
FEF Forced expiratory flow from 25–75% of FVC
M, F Male, female
[1] O_3 concentration within purified air
[2] O_3 concentration within ambient air
[3] $FEV_{0.75}$

Source: Lippman, 1989a

Susceptibility to O_3 exposure does not appear to vary by ethnic (Seal *et al.*, 1993) or demographic characteristics (Messineo and Adams, 1990). Dietary antioxidant levels have been shown to modulate the response to O_3 exposure in animals (Slade *et al.*, 1985; Elsayed *et al.*, 1988), but data for humans are still sparse. In controlled clinical studies, asthmatics do not appear to be more sensitive to O_3 as shown by their FEV_1 response (Koenig *et al.*, 1989a). However, short-term follow-up of asthmatic subjects (panel study) exposed to O_3 have shown a decrease in lung function and an increase in respiratory symptoms among adults (Holguin *et al.*, 1984; Lebowitz *et al.*, 1987) and children (Kryzanowski *et al.*, 1992; Hoek *et al.*, 1993; Neas *et al.*, 1995; Ostro *et al.*, 1995; Romieu *et al.*, 1997). Other studies were unable to find any adverse effect on respiratory health (Vedal *et al.*, 1987; Roemer *et al.*, 1993). However, in these studies, ambient O_3 concentrations were low and this could partly explain the results. Ozone may exacerbate asthma by facilitating the entry of allergens or because of the inflammation it induces. There is some evidence that O_3 may act synergistically with other pollutants, such as sulphate and NO_2 (Kleinman *et al.*, 1989b). Koenig *et al.* (1989a) showed that inhaling low concentrations of O_3 may potentiate the bronchial hyper-responsiveness of people with asthma to SO_2 exposure. A similar potentiating effect has been observed with exposures to O_3 and sensitivity to NO_2 (Hazucha *et al.*, 1994).

The long-term exposure effects of O_3 are still unclear, but there is good reason for concern that repeated exposure could lead to chronic impairment of lung development and functions. Animal studies have demonstrated progressive epithelial damage and inflammatory changes that appear to be cumulative and persistent, even in animals that have adapted to exposure by modifying their respiratory mechanisms (Tepper *et al.*, 1989) at O_3 concentrations slightly higher than those that produce effects in humans. Furthermore, for some chronic effects, intermittent exposures can produce greater effects than continuous exposure regimes that result in a higher cumulative exposure. These results suggest that disease pathogenesis depends on the effects produced by lung defensive responses to the direct damage caused by O_3 to epithelial cells, as well as on the direct effects themselves (Lippmann, 1989a).

Epidemiological studies on populations living in Southern California suggest that chronic oxidant exposures affect baseline respiratory function. Comparing two communities from this area, Detels *et al.* (1987) found that baseline lung functions were lower and that there was a greater rate of decline in lung function in the high oxidant community over 5 years. However, this study has been criticised for several reasons (poor exposure measurements, lack of adjustment for potential confounding factors such as indoor air pollution, and occupational exposure).

Kilburn *et al.* (1992), comparing the lung functions of school children from Los Angeles and Houston, observed that children from Los Angeles had 6 per cent lower baseline values for FEV_1 and 15 per cent lower values for FEF_{25-75} than children from Houston. Aerosol administration of metaproterenol to Los Angeles children improved FEV_1 by 1 per cent and FEF_{25-75} by 6.6 per cent, but expiratory flows were still below the values of Houston children, suggesting that impairment was not reversible.

Repeated measurements of lung functions among 106 Mexican-American children from Los Angeles showed that FEV_1 and FEF_{25-75} were respectively 2 per cent and 7 per cent lower than the predicted value in 1987 compared with 1984. The FVC remained unchanged. The authors concluded that the worsening of airway obstruction in these children was probably due to air pollution (Kilburn *et al.*, 1992). Data from the National Health and Nutrition Examination Survey (NHANES II) suggested that ambient O_3 concentrations were associated with a loss of lung function when the annual average concentration was above 80 $\mu g\ m^{-3}$ (0.04 ppm) (Schwartz, 1989).

Euler *et al.* (1988) evaluated the risk of chronic obstructive pulmonary disease due to long-term exposure to ambient levels of total oxidants and NO_2 in a cohort of 7,445 Seventh Day Adventists non-smokers, who had resided in California for at least 11 years and were also aged at least 25 years. The results suggested a significant association between chronic symptoms and total oxidants exceeding 200 $\mu g\ m^{-3}$ (0.10 ppm). However, when cumulative exposure to SPM was considered as a confounding variable in the model, only SPM exposure above 200 $\mu g\ m^{-3}$ showed statistical significance. Other studies conducted among Seventh Day Adventists suggested that O_3 exposure could lead to an increase in respiratory cancer (Abbey *et al.*, 1993). This result should be interpreted with caution because of small number of cases studied. Sherwin and Richter (1990) conducted an autopsy study among 107 young non-smoker adults aged between 14–25 years, and who died of non-respiratory traumatic causes in Los Angeles County. They found that 29 of them had lungs with severe respiratory bronchiolitis of the kind first described in young smokers. Moderate changes were present in a further 51 cases.

In summary, effects which have been associated with hourly average oxidant concentrations beginning at about 200 $\mu g\ m^{-3}$ (0.10 ppm), include eye, nose and throat irritation, cough, throat dryness, thoracic pain, increased mucous production, rales, chest tightness, substernal pain, lassitude, malaise and nausea.

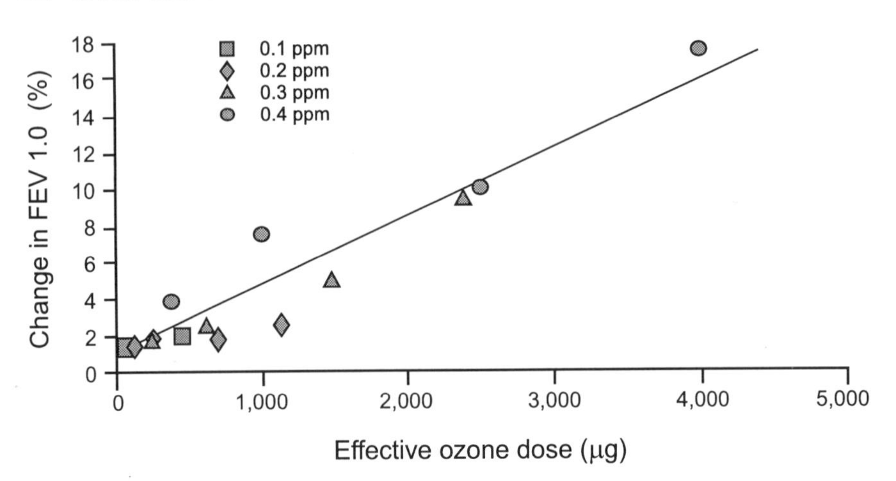

Figure 2.3 The effect of ozone on pulmonary function (After Kleinman *et al.*, 1989a)

The transient effects of O_3 seem to be more closely related to cumulative daily exposure than to 1-hour peak concentrations. Several studies provide sufficient information to allow a quantitative evaluation of the potential impact of short-term exposure to O_3 on population subgroups (WHO, 1995a). Kleinman *et al.* (1989a), summarising data from different studies, derived a dose-response relationship between change in FEV1 and effective dose of O_3 (product of concentration of O_3 × time × ventilation rate, see Figure 2.3). These authors also calculated dose-response functions to determine the change in the percentage of the population affected by specific symptoms according to O_3 ambient level. The model fitted well data derived from clinical studies. Finally, the work of Ostro (1989) can be used to determine the change in restricted activity days associated with changes in ambient O_3.

Given the high degree of correlation between the 1-h and 8-h O_3 concentrations in field studies, an improvement in health risk associated with decreasing 1-h or 8-h O_3 levels should be almost identical. Consequently, Table 2.4 presents an overview of the health outcomes associated with the change in peak daily ambient O_3 concentrations observed in epidemiological studies (WHO, 1995a). However, these models can only provide an approximation to the health outcomes because the effects of O_3 may be enhanced by the presence of other environmental variables such as acid aerosols. The effects of long-term chronic exposure to O_3 remain poorly defined, but recent epidemiological and animal inhalation studies suggest that current ambient levels (close to 240 μg m^{-3} or 0.12 ppm) are sufficient to cause premature lung ageing (Lippmann, 1989b, 1993). Animal data suggest a morphological change in terminal and respiratory bronchioles with an associated restrictive

Table 2.4 Health outcomes associated with changes in peak daily ambient ozone concentrations in epidemiological studies

Health outcome	Changes in 1-h O_3 ($\mu g\ m^{-3}$)	Changes in 8-h O_3 ($\mu g\ m^{-3}$)
Symptom exacerbation among healthy children and adults or asthmatics during normal activity		
25% increase	200	100
50% increase	400	200
100% increase	800	300
Hospital admissions for respiratory conditions		
5% increase	30	25
10% increase	60	50
20% increase	120	100

Source: WHO, 1995a

impairment in airflow. More research is needed on the chronic effects of O_3 on lung structure, disease pathogenesis and interaction with other environmental factors.

The WHO proposed guideline for O_3 is an 8-h (moving average) value of 120 $\mu g\ m^{-3}$ (0.06 ppm) (WHO, 1995a). Although, previous recommendations have included a 1-h guideline value of 150–200 $\mu g\ m^{-3}$ (0.076–0.1 ppm), it is believed that the 8-h guideline would protect against 1-h exposure in this range, consequently a 1-h guideline is not necessary (WHO, 1995a).

2.4 Sulphur dioxide, suspended particulate matter and acid aerosols

Sulphur dioxide and suspended particulate matter, are the primary products of fossil fuel combustion processes. Secondary particles, including acid aerosols, are formed by atmospheric chemical reactions. The chemical properties of atmospheric particles are diverse, reflecting the complexity of the sources (Spengler *et al.*, 1990).

Historically, SO_2 and SPM have been associated, in terms of sources and environmental concentrations, with the use of coal for domestic heating and industrial purposes. However, there has been a great increase in vehicular traffic in the past 20 years; SPM from motor vehicle emissions and secondary aerosols arising from atmospheric reactions in the gaseous and liquid phases have led to relatively high concentrations of particles which are less than 0.2 μm in aerodynamic diameter (Pope *et al.*, 1995; WHO, 1995a). Recent findings on the adverse health effects of SO_2, SPM and acid aerosols at concentrations close to the current guidelines for these contaminants have led to discussions on the concept of "safe level".

Inhaled SO_2 is highly soluble in the aqueous surfaces of the respiratory tract. It is therefore absorbed in the nose and the upper airways where it exerts an irritant effect; very little SO_2 reaches the lungs. In addition to irritation of the upper airways, high concentrations can cause laryngotracheal and pulmonary oedema. From the respiratory tract, SO_2 enters the blood. Elimination occurs mostly by the urinary route (after biotransformation to sulphate in the liver).

Airborne particulate matter represents a complex mixture of organic and inorganic substances. Mass and composition tend to divide particulate matter into two principal groups: coarse particles larger than 2.5 μm in aerodynamic diameter and fine particles smaller than 2.5 μm. The smaller particles contain the secondarily-formed aerosols (gas to particle conversion), combustion particles and recondensed organic and metal vapours (WHO, 1987c). The biological effects of inhaled particles are determined by the physical and chemical properties of the particles, the sites of deposition, and the mechanisms by which the particles injure the lung. The deposition of particulate matter depends mainly on the breathing pattern and the particle size. With normal nasal breathing, larger particles or aerodynamic diameter between 10 and 100 μm are mainly deposited in the extrathoracic part of the respiratory tract and most of the particles in the range 5–10 μm are deposited in the proximity of the fine airways. With mouth breathing the proportion of tracheobronchial and pulmonary deposition increases. Because of their small size (fine particles), acid ambient aerosols tend to be deposited in the distal lung airway and airspace. Some neutralisation of the droplets can occur before deposition due to normal excretion of endogenous ammonia into the airways. Deposited free H^+ reacts with components of the mucus of the respiratory tract changing its viscosity; the non-reacting part diffuses into the tissues (WHO, 1987c). Current concepts of particle toxicity emphasise the role of particle acidity and the induction of inflammation at the sites of injury (Bascom *et al.*, 1996).

Controlled human studies have shown acute effects of SO_2 on pulmonary functions (reduction in FEV_1 and increase in specific airway resistance) among asthmatics as well as an increase in symptoms, such as wheezing and shortness of breath (WHO, 1995a). Horstmann *et al.* (1988) have shown that when asthmatic subjects exercise in air containing SO_2, symptomatic bronchoconstriction can develop within minutes when concentrations are close to 715 μg m^{-3} (0.25 ppm). Linn *et al.* (1987) report a 10 per cent decrease in FEV_1 after 15 minutes exposure to 1,144 μg m^{-3} (0.4 ppm), and a reduction of 15 per cent after exposure to 1,716 μg m^{-3} (0.6 ppm) among moderate and severe asthmatics. It is well established in animal experiments and controlled human studies that acid aerosols have a deleterious effect on respiratory health. Lippmann (1989c) reviewed the state of knowledge on these effects. Alteration of lung functions, particularly increase in pulmonary flow resistance, occurs after acute exposure. Sulphuric acid (H_2SO_4) appears to be more potent than any of the sulphate salts in terms of increased airway irritability (WHO, 1987c). Acid aerosols have also been shown to modify particle clearance, although the mechanism is not yet well established (Folinsbee, 1989). The lowest demonstrated effect concentration for sulphuric acid was 100 $\mu g\ m^{-3}$ with mouthpiece inhalation and intermittent exercise (Hackney *et al.*, 1989).

Although controlled exposures to different concentrations have shown different effects on respiratory functions, epidemiological studies have provided much of the information concerning the effects of exposure to realistic concentrations of SO_2, SPM and acid aerosols. Variations in the 24 hour average concentrations of SO_2 and SPM have been associated with increased mortality and morbidity, and reductions in lung function. Recent findings have been extensively reviewed by several authors (Dockery and Pope, 1994; Schwartz, 1994a; Pope *et al.*, 1995; Samet *et al.*, 1995).

During the first half of this century, episodes of marked air stagnation have resulted in well-documented excess mortality in areas where fossil-fuel combustion resulted in very high levels of SO_2 and SPM (US EPA, 1982b). In one notable episode in Donora, Pennsylvania in October 1948, 43 per cent of the population of approximately 10,000 were adversely affected. A similar event occurred later in London where concentrations of SO_2 and smoke rose above 500 $\mu g\ m^{-3}$. The people primarily affected were those with pre-existing heart and lung disease and the elderly, although closer examination of these data, suggested that children under 5 years old were also severely affected (Logan, 1953; Ministry of Health, 1954).

Following these major episodes, attention was turned to studies on more moderate day-to-day variations in mortality within large cities, in relation to pollutants. These acute mortality studies of SO_2 and particulate matter suggested a dose-response relationship between 24 hour levels of these pollutants and excess mortality, particularly at concentrations over 500 $\mu g\ m^{-3}$. However, new analysis of the London data for the winters of 1958–59 to 1971–72, controlling for important confounding variables such as temperature and humidity, indicates the absence of a threshold level for the adverse effect of British smoke; however a statistically significant pollutant effect on mortality was observed for $< 150\ \mu g\ m^{-3}$ (Ostro, 1994). Similar results have been reported by other authors (Schwartz and Marcus, 1990; Schwartz, 1991). A re-analysis of the London smog mortality data in relation

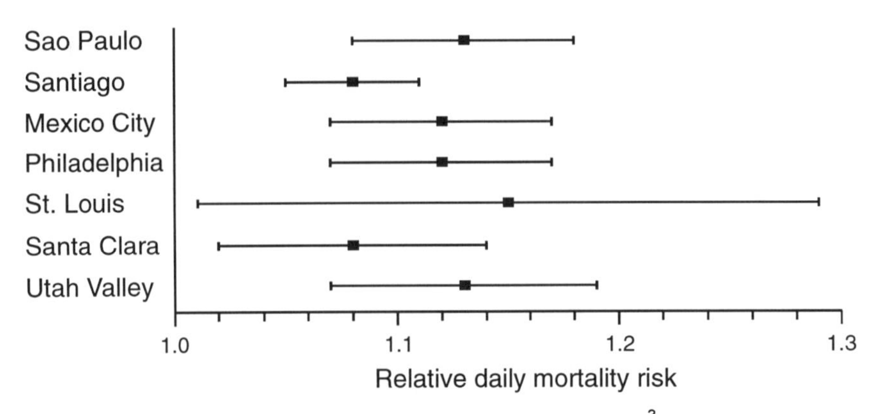

Figure 2.4 Relative risk of daily mortality related to a 100 μg m^{-3} increase in PM_{10} in different cities (After Romieu and Borja-Aburto, 1997)

to exposure to acid aerosols suggested that the SO_4^{2-} was the component of greatest health significance (Thurston *et al*., 1989).

In the past few years, several studies have used time-series analysis to look at the relative change in the rate of mortality associated with changes in air quality variables. In these studies weather conditions and other potential confounding factors were accounted for. Although various measurements of particulate pollution were used, the result of most studies suggested that a 10 μg m^{-3} increase in PM_{10} (mean aerodynamic particle diameter of 10 μm) was associated with an increase in daily mortality equal to 0.5–1.5 per cent (Pope *et al*., 1995) and the absence of a threshold level.

Some studies also provide a breakdown of mortality by causes of death. A summary estimate of these studies suggested that an increase of 10 μg m^{-3} in PM_{10} (24-h average) was related to a 3.4 per cent increase in respiratory mortality (range 2–8 per cent) and a 1.4 per cent increase in cardiovascular mortality (range 0.8–1.8 per cent) (Dockery and Pope, 1994). A detailed examination of cause of deaths showed the relative increase was higher for chronic obstructive lung diseases and pneumonia. Also, almost all cardiovascular deaths on days with high particulate air pollution have a respiratory cause as a contributing factor (Schwartz, 1994b). It is important to note the consistency of the results across locations.

Similar results have been reported in several European countries (Katsouyanni *et al.,* 1995; Touloumi and Katsouyanni, 1995; Zmirou *et al*., 1995) as well as in Latin American countries (Romieu and Borja-Aburto, 1997). Figure 2.4 presents results from some of these studies.

The specific biological mechanism for the acute increase in mortality is still not clear (Utell, 1994). Whether the particulate or the SO_2 component of combustion mixtures is responsible for adverse health effects is still being debated. Results from studies conducted in Utah Valley and Santa Clara, California, where SO_2 concentrations are low, tend to favour an adverse effect of particulate matter.

It has been proposed that the pollutant really at fault could be a constituent of particulate matter such as combustion particles, sulphate particles, of fine- or ultra-fine particles (Pope *et al.*, 1995). Also, some groups of individuals seem to be more susceptible to particulate matter and SO_2 pollution. These include persons with heart or lung conditions, in particular chronic obstructive pulmonary diseases (COPD), asthma, and ischemic heart disease, particularly infants and elderly persons and possibly people with acute respiratory infections (Bascom *et al.*, 1996). It has been suggested that exposure to air pollution would decrease the life expectancy of more susceptible persons only insignificantly. However, a mortality effect from long-term or chronic exposure to particulate air pollution has also been observed in several cross-sectional studies.

Short-term peak concentrations of SO_2 and particulate matter may also increase morbidity, especially in individuals with higher sensitivity than the general population, such as those with asthma and chronic bronchitis. Various studies have evaluated acute morbidity effects of particulate pollution by examining short-term temporal associations between hospital admissions or emergency visits as well as changes in lung functions and in respiratory symptoms. Increased hospital admissions due to respiratory diseases, including asthma, have been observed in Southern Ontario in relation to sulphate concentrations (Bates and Sizto, 1987; Burnett *et al.*, 1994). Peaks of acid aerosols observed in Canada during the summer could have been responsible for the association observed between sulphate concentrations and hospital admissions, to the extent that sulphate constitutes a surrogate variable for sulphuric acid aerosols (Bates and Sizto, 1987). Similar results have been observed in other parts of the USA. (Dockery and Pope, 1994). Taken together, these studies found an increase of hospital admissions for all respiratory diagnoses ranging from 0.8 per cent to 3.4 per cent (weighted mean, 0.8 per cent) for each 10 $\mu g\ m^{-3}$ increase in daily mean PM_{10}. In studies on respiratory-related emergency visits, an increase of 0.5 per cent to 3.4 per cent for each increase of 10 $\mu g\ m^{-3}$ in PM_{10} has been reported; (Samet *et al.*, 1981; Schwartz *et al.*, 1993; Sunyer *et al.*, 1993).

Asthmatic subjects seem to be more susceptible to SO_2 and particle pollution, and several studies have evaluated the impact of PM_{10} on the occurrence of asthma attacks and on the frequency of use of bronchodilators. The incidence of asthma attacks increased from 1.1 to 11.5 per cent (weighted mean,

3 per cent) with an increase of 10 μg m^{-3} in PM_{10} daily average, and bronchodilator use increased from 2.3 per cent to 12 per cent (weighted mean, 2.9 per cent) (Dockery and Pope, 1994). Using data from the Health Interview Survey (HIS) collected over a 6-year period (1976–81), Ostro (1989) calculated that an increase in the annual mean of 1 μg m^{-3} in fine particulate matter (2.5 μm) could be associated with a 3.2 per cent increase in acute respiratory diseases in adults aged 18–65 years.

In some studies, observed deviations in the lung function of children have been associated with short-term fluctuations in particulate concentration (WHO, 1987c). From data collected by Dockery *et al.* (1982) during air pollution episodes it can be calculated that in the most sensitive children (approximately 25 per cent of the study population), a deficit in lung function at least four times greater than in those of average sensitivity was observed, corresponding to a decrease in FEV_1 of 0.39 ml per each increase of 1 μg m^{-3} of exposure to SPM. The minimum concentration for effect was judged to be 180 μg m^{-3} SPM. However, more recent panel studies have suggested the absence of a threshold value in the adverse health effect of particulate matter (WHO, 1995a).

In a study conducted among asthmatics in Utah Valley, PM_{10} concentrations were associated with reductions in PEFR and increased symptoms and medication use (Pope *et al.*, 1991). Aerosol acidity was the below detection limit. Decreases in lung function parameters related to particle exposure have also been observed in studies conducted in the Netherlands (Hoek and Brunekreef, 1993; Roemer *et al.*, 1993), Germany (Peters *et al.*, 1995) and Mexico (Romieu *et al.*, 1996).

Studies conducted in Union Town showed a decrease in lung function associated with aerosol acidity as well as with PM_{10} (Neas *et al.*, 1992, 1995). Raizenne *et al.* (1989) reported a 3.5–7 per cent decrease for FEV_1 and PEFR associated with air pollution episodes with maximum O_3 concentrations of 286 μg m^{-3} (0.143 ppm) and 47.7 μg m^{-3} of H_2SO_4. Most of the studies suggested that a 10 μg m^{-3} increase in respirable particles resulted in less than 1 per cent decline in lung function (Pope *et al.*, 1995). In addition to declines in lung function, many of these studies observed increases in respiratory symptoms. A 10 μg m^{-3} increase in PM_{10} was typically associated with a 1–10 per cent increase in respiratory symptoms such as cough, combined lower respiratory symptoms and asthma attacks. These effects were also observed at comparable PM_{10} concentrations near, or even below, 150 μg m^{-3} (Pope *et al.*, 1995).

Asthmatics are more sensitive in terms of changes in lung function than healthy people, and vigorous exercise potentiates health effects at a given concentration of SO_2 or SPM. A preliminary analysis of data from Koenig *et al.* (1989b) collected among allergic children, showed that exposure to H_2SO_4 alone, or in association with SO_2, caused significant changes in lung functions, whereas exposure to relatively clean air or SO_2 in the absence of acid aerosols did not. Recent studies by Ostro *et al.* (1991) in Denver have produced evidence that aerosol H^+ levels were associated with worsening asthma. However, the acute health effects observed in these studies were transient and did not give any indications of the long-term effect of exposure to SO_2 and particulate air pollution.

Studies on long-term health effects have related annual means of SO_2 and suspended particulate matter to mortality and morbidity (Samet *et al.*, 1995). Ecological studies on the relationship between SO_2 and particulate levels and mortality from cardio-respiratory diseases, have usually indicated that this complex association of SO_2 and SPM typically accounts for approximately 4 per cent of the variation in death rates between cities (WHO, 1987c). However, these cross-sectional studies have an inherent limitation due to their design. Many factors, such as differences in smoking habits, occupation or social conditions may contribute to the disparities in death rates attributed to SO_2 and particulates. Nevertheless, the results of studies carried out in different parts of the world imply a relatively consistent association between long-term residence in more polluted communities and increased mortality rates. Evans *et al.* (1984) summarised data from 23 original cross-sectional mortality studies and derived a dose-response function to assess the impact on mortality of change in particulate levels.

More recently, Dockery *et al.* (1994) and Pope *et al.* (1995) have determined cross-sectional differences in mortality among cohorts of adults in the USA. The results from the first study suggested that a 10 $\mu g\ m^{-3}$ increase in average PM_{10} exposure was associated with an increase in chronic mortality equal to 9 per cent (odds ratio of 1.09 with 95 per cent confidence interval between 1.03 and 1.15). This estimate reached 13 per cent for fine particles ($PM_{2.5}$), with a 95 per cent confidence interval between 1.04 and 1.23, and 36 per cent for sulphates with a 95 per cent confidence interval between 1.10 and 1.62. In the second study (Pope *et al.*, 1995) an increase of 7 per cent was observed for total mortality for each 10 $\mu g\ m^{-3}$ increase in $PM_{2.5}$. The strongest association was observed with cardiopulmonary disease and lung cancer deaths (Pope *et al.*, 1995).

Community-based studies conducted among adults and children have indicated a detectable increase in the frequency of respiratory symptoms and illnesses in communities where annual mean concentrations of both black smoke and SO_2 exceed 100 $\mu g\ m^{-3}$ (WHO, 1987c). However, recent studies related to industrial sources of SO_2 or to the changed urban mixture have shown adverse effects below this level. A major difficulty in interpretation is that long-term effects are liable to be affected not only by current conditions, but also by qualitatively and quantitatively different levels of pollution in earlier years (WHO, 1995a).

Several studies have investigated the relationship between respiratory illness and symptom rates, and pollutant levels. In a study conducted among adult British residents, Lambert and Reid (1970) found a higher prevalence rate of chronic bronchitis in residents of areas with heavier SO_2 and particulate air pollution, independent of cigarette smoking, although an interacting effect was also present. Results from other correlation studies have tended to suggest that areas with higher pollution levels are associated with higher levels of chronic bronchitis, although the interpretation of these results is limited by the crude exposure measurement.

In a prospective study conducted in the Netherlands (Van de Lende *et al.*, 1981) in two communities over a 12-year period, residents of the community with higher pollution levels had higher rates of lung function decline, implying that exposure to high ambient pollution during adulthood may be a risk for COPD. In a study conducted among a preadolescent population aged 6–9 years living in six USA cities (Ware *et al.*, 1986), frequency of chronic cough was significantly associated with the annual average concentration of these air pollutants (SPM, total sulphates (TSO_4) and SO_2) during the year preceding the examination (Figure 2.5). The maximum concentrations observed were 114 $\mu g\ m^{-3}$ for SPM, 68 $\mu g\ m^{-3}$ for SO_2 and 18 $\mu g\ m^{-3}$ for TSO_4. The rates of bronchitis and composite measure of respiratory illness were significantly associated with average particulate concentration. Similar results have been confirmed in a second cross-sectional survey of the same population (Dockery *et al.*, 1989). A subject with asthma and permanent wheeze experienced a higher rate of pulmonary symptoms in relation to increased pollutants. There was no evidence of impaired lung function associated with pollutant levels.

Other studies conducted over a period of years have shown an association between the magnitude of lung function changes and the level of pollution (WHO, 1987c); however there were no firm conclusions because of the lack of accurate exposure measurement. A study conducted among Seven-Day Adventists living in California reported an association between cumulative exposure to SO_2 and SPM and symptoms of chronic pulmonary disease defined as chronic cough or phlegm, a doctor's diagnosis of asthma with wheeze or a diagnosis of emphysema with shortness of breath (Euler *et al.*, 1987). Acid aerosols seem to act synergistically with O_3. Stern *et al.* (1989)

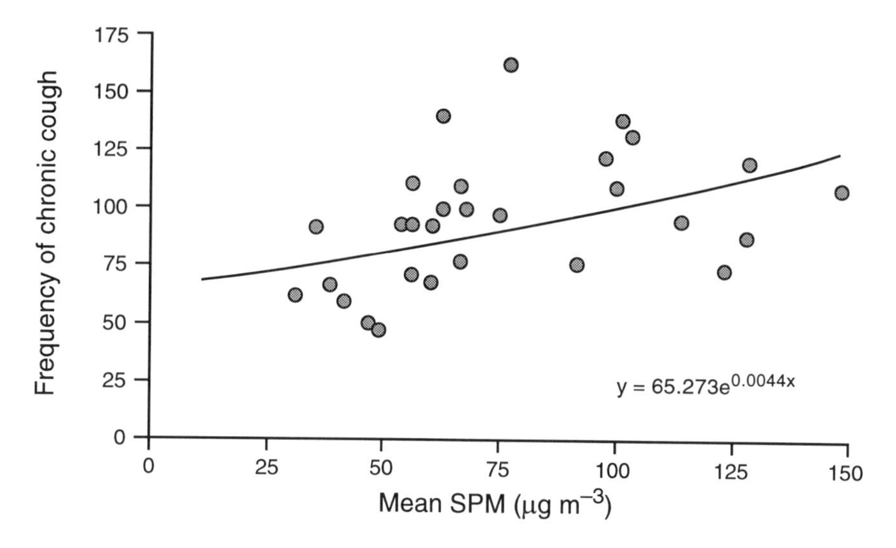

Figure 2.5 The relationship between adjusted frequency of chronic cough and total particulate levels (After Ware *et al.*, 1986)

investigated respiratory health effects associated with ambient sulphates and O_3 in two rural Canadian communities in 1983–84. Respiratory health was assessed by the measurement of lung functions and by evaluation of the child's respiratory symptoms and illness using a parent-completed questionnaire. There was a significant difference in the level of SO_2, SO_4^{2-} and NO_2 between the two communities. Children living in the community with the highest levels of contaminants had a significant decrease in lung function parameters (2 per cent for FVC and 1.7 per cent for FEV_1).

There is some growing evidence that chronic exposure to smoke may play an important role in the genesis of chronic lung disease. In developing countries, prevalence rates of chronic bronchitis often appear to be much higher than in industrialised countries, and with sex ratios tending to 1, which cannot be explained solely by cigarette smoking (Bumgartner and Speizer, 1991). Although exposure to multiple risk in developing countries may be much higher than in developed countries, these results suggest that exposure to indoor smoke pollution which is much more common among women, may largely account for the differences.

Non-neoplastic and neoplastic effects on the lung from exposure to diesel engine exhaust have recently been reviewed by a WHO Task Group on Environmental Health Criteria for diesel fuel and exhaust emissions (WHO, 1996a). Non-neoplastic effects of diesel exhaust include mucous membrane irritation, headache and light-headedness. Diesel-bus garage workers reported significantly more incidents of cough, itchy or burning eyes, headache and wheezing. Several cases of persistent asthma and asthma attacks have also been reported after acute exposure. Four studies of occupationally exposed individuals with long, well-defined exposure and follow-up (> 20 years) investigated the prevalence of lung cancer (Garshick *et al.*, 1987, 1988; Gustavsson *et al.*, 1990; Emmelin *et al.*, 1993). All four studies showed an increased risk of lung cancer with exposure to diesel exhaust. The relative risks reported ranged between 1.3 and 2.7. The point estimates were, however, imprecise and had wide confidence intervals including a relative risk of 1.0 as a lower limit of the 90 per cent confidence intervals. Other studies supported these results with relative risks in the range 1.2–1.9 but did not always achieve statistical significance (WHO, 1996b).

In summary, recent findings suggest that there is no threshold for the adverse health effects of SPM that may occur when ambient levels are lower than the older WHO guideline for respirable particles (WHO, 1987c) or the current US EPA standard for PM_{10} (US EPA, 1982b). It seems that the role of secondary products such as acid sulphate are strongly involved in the adverse effects of the SO_2–particulate matter complex (Spengler *et al.*, 1990). Findings of several studies have been used for quantitative evaluations of the health impact of particulate matter and have also been used for risk assessment (Kleinman *et al.*, 1989a; Ostro, 1990; Romieu *et al.*, 1990). Recent findings (Table 2.5) provide more insight into the amplitude of the effect and have suggested that at low concentrations of 24-h exposure (defined as 0–200 $\mu g\ m^{-3}$ PM_{10}), the exposure–response curve probably fits a straight line reasonably well. However, some studies conducted in an area with a higher level of air pollution (several hundreds of $\mu g\ m^{-3}$ PM_{10}) suggest a curvilinear relationship with the slope becoming shallower at higher ambient concentrations. Estimates of the magnitude of effect occurring at low levels of exposure should, therefore not be used to extrapolate to higher concentrations outside the range of exposures that existed in most of the recent acute studies (WHO, 1995a). Particle composition, and size distribution within PM_{10} fraction, as well as synergy with other pollutants, are important factors. Also, it is not clear whether long-term effects can be related simply to annual mean values or to repeated exposure to peak values. Better specification of these effects is needed, especially for the genesis and evolution of chronic pulmonary diseases.

In many developing countries, people are exposed to PM_{10} levels that are greatly in excess of US EPA standards and there is great concern for the impact of these levels on the health of population exposed. Although there is an urgent need to implement control strategies to decrease air pollution levels, further studies are needed: to understand better the effects of

Table 2.5 Health effects associated with short-term exposure to 10 μg m^{-3} increases in PM_{10}

Health effect	Mean % changes for each 10 μg m^3 increase in PM_{10}	Range of % changes for each 10 μg m^{-3} increase in PM_{10}
Daily mortality		
Total	1.0	0.5–1.5
Cardiovascular	1.4	0.8–1.8
Respiratory	3.4	1.5–3.7
Morbidity		
Hospital admission for respiratory condition	1.1	0.8–3.4
Emergency visits for respiratory conditions	1.0	0.5–4
Symptom exacerbation among asthmatics	3.0	1.1–11.5
Changes in peak expiratory flow	0.08	0.04–0.25

Source: Dockery and Pope, 1994; Pope *et al.*, 1995

particulate air pollution on health at levels exceeding that observed in western countries, to evaluate the impact of intervention strategies, to determine the chronic effect of particulate exposure on health mortality and morbidity and the interactive factors that may modulate these effects.

The WHO guidelines for ambient concentrations of SO_2 are: 500 μg m^{-3} (0.175 ppm) for 10 minutes, 125 μg m^{-3} (0.044 ppm) for the 24-h average and 50 μg m^{-3} (0.017 ppm) for the annual average (WHO, 1995a). Given the recent epidemiological findings that suggest the absence of a threshold for effects on morbidity and mortality of particulate matter, no guidelines values have been set by WHO for this pollutant but risk considerations have been recommended.

The norms provided by the US EPA for total SPM are: 75 μg m^{-3} for the annual average and 260 μg m^{-3} for the 24-h (US EPA, 1982b). Since 1987, the US EPA has restricted the national air quality standards to inhalable particles less than 10 μm aerodynamic diameter to 50 μg m^{-3} for the annual average and 150 μg m^{-3} for the 24-h average (US EPA, 1987). It is important to mention that these norms are actually under revision by the US EPA.

2.5 Carbon monoxide

Carbon monoxide is rapidly absorbed in the lungs and is taken up in the blood where it binds to haemoglobin (Hb) with the formation of carboxyhaemoglobin (COHb), which impairs the oxygen carrying capacity of blood.

Table 2.6 Predicted carboxyhaemoglobin levels for people engaged in different types of work in different concentrations of carbon monoxide

CO concentration (ppm)	CO concentration (mg m^{-3})	Exposure time	Predicted COHb level for those engaged in: Sedentary work	Light work	Heavy work
100	115	15 min	1.2	2.0	2.8
50	57	30 min	1.1	1.9	2.6
25	29	1 h	1.1	1.7	2.2
10	11.5	8 h	1.5	1.7	1.7

Source: WHO,1987d

The dissociation of oxyhaemoglobin is also altered due to the presence of COHb in the blood thereby further impairing the oxygen supply to body tissues (the affinity of haemoglobin for CO is about 240 times that of oxygen). Carbon monoxide is also bound to myoglobin (MbCO) in cardiac and skeletal muscle which will limit the rate of O_2 uptake by these tissues and impair O_2 delivery to intercellular contractile processes (Agostoni *et al.*, 1980). The main factors conditioning the uptake of CO are its concentration in the inhaled air, the endogenous production of CO, the intensity of physical effort, body size, the condition of the lungs, and the barometric pressure. Table 2.6 presents the expected COHb levels after exposure to CO concentrations between 11.5 and 115 mg m^{-3} during different types of physical activity. In absence of CO exposure, COHb concentrations are approximately 0.5 per cent; one-pack per day cigarette smokers may achieve COHb saturation of 4–7 per cent (WHO, 1979).

The main effect of CO is to decrease the oxygen transport to the tissues. Organs which are dependent on a large oxygen supply are the most at risk, particularly the heart and the central nervous system, as well as the foetus. Four types of health effects are reported to be associated with CO exposure: neurobehavioral effects, cardiovascular effects, fibrinolysis effects and perinatal effects. Carbon monoxide leads to a decreased oxygen uptake capacity with a resultant decreased work capacity under maximum exercise conditions. According to available data, the COHb level required to induce these effects is approximately 5 per cent (WHO, 1979). Some authors (Beard and Wertheim, 1967) have reported an impairment in the ability to judge correctly slight differences in given tasks in successive short-time intervals at lower COHb levels of 3.2 to 4.2 per cent. At this level, subjects may miss signals they would not have missed when starting a task.

Subjects with previous cardiovascular disease (chronic angina patients) seem to be the group most sensitive to CO exposure. In a recent study, Allred *et al.* (1989) investigated the effects of CO exposure on myocardial ischaemia during exercise in 63 men with documented coronary artery disease. Results showed a decreasing dose–response relationship between the length of time to the onset of angina and COHb level. The length of time to the onset of angina was reduced by 4.2 per cent for COHb levels of 2 per cent and by 7.1 per cent at 3.9 per cent COHb level. This study shows that COHb levels as low as 2 per cent can exacerbate myocardial ischemia during exercise in subjects with coronary artery disease. Similar effects have been demonstrated in patients with intermittent claudication from peripheral vascular disease (Aronow *et al.*, 1974).

A retrospective cohort study conducted among bridge and tunnel officers ($n = 5{,}529$) exposed to CO showed a 35 per cent excess risk of arteriosclerotic heart disease mortality among tunnel officers when compared with the New York City population. Two factors contributed to this excess risk: the exposure (current) to CO of tunnel officers and the movement into a critical higher age group. There was a reduction in mortality signifying a decrease in exposure (more ventilation in the tunnel) (Stern *et al.*, 1988). Table 2.7 summarises studies relating human health effects to different low-level exposures to CO. The classic symptoms of CO poisoning are headache and dizziness at COHb levels between 10 and 30 per cent and severe headache, cardiovascular symptoms and malaise at above about 30 per cent. Above about 40 per cent there is considerable risk of coma and death.

Carbon monoxide exposure may also affect the foetus directly through oxygen deficit without elevation of COHb level in the foetal blood. During exposure to high CO levels, the mother's Hb gives up its oxygen, less readily with a consequent lowering of the oxygen pressure in the placenta, and hence also in the foetal blood. Research has mainly focused on the effects of cigarette smoking during pregnancy. The main effects are reduced birth weight (Hebel *et al.*, 1988; Ash *et al.*, 1989; Mathai *et al.*, 1990) and retarded postnatal development (Campbell *et al.*, 1988). Impairment of neurobehavioral development has also been related to maternal smoking during pregnancy (US EPA, 1983). Ambient CO exposure has been related to low birth weight in a case-control study conducted in Denver, USA (Alderman *et al.*, 1987). Mothers who lived in neighbourhoods with a mean CO concentration below 3.4 mg m^{-3} (3 ppm) during the last trimester of their pregnancy had a 50 per cent increase in the risk of having an infant with low birth weight when compared with a control group unexposed to CO (odds ratio 1.5 with a 95 per cent confidence interval between 0.7 and 3.5). However, in this study personal smoking history was not recorded, which may have biased the results.

Table 2.7 Human health effects associated with low-level carbon monoxide exposure

Carboxyhaemoglobin concentration (%)	Lowest observed effect level (LOEL)
2.3–4.3	Statistically significant decrease (3–7%) in the relation between work time and exhaustion in exercising young healthy men
2.9–4.5	Statistically significant decrease in exercise capacity (i.e. shortened duration of exercise before onset of pain) in patients with angina pectoris and increase in duration of angina attacks
5–5.5	Statistically significant decrease in maximal oxygen consumption and exercise time in young healthy men during strenuous exercise
< 5.0	No statistically significant vigilance decrements after exposure to CO
5–7.6	Statistically significant impairment of vigilance tasks in healthy experimental subjects
5–17	Statistically significant diminution of visual perception, manual dexterity, ability to learn, or performance in complex sensorimotor tasks (e.g. driving)
7–20	Statistically significant decrease in maximal oxygen consumption during strenuous exercise in young healthy men

Source: WHO, 1987d

In summary, average COHb levels in the general population are around 1.2–1.5 per cent (in cigarette smokers around 3–4 per cent). Below 10 per cent COHb, it is mainly cardiovascular and neurobehavioral effects that have been evaluated. The aggravation of symptoms in angina pectoris patients, which is a major public health concern, may occur at levels as low as 2 per cent COHb. Decreased work capacity and neurobehavioral function have mostly been observed at around 5 per cent COHb. Low birth weight has been related to cigarette smoking during pregnancy with the hypothesis that an increase in a mother's COHb could have a role in this adverse effect, however there is no estimation of the impact of specific COHb level on the decrease in birth weight. Based on ambient CO concentration, time of exposure and physical activity type, the expected level of COHb can be derived (Table 2.6) and used to determine the health impact on the population of specific CO exposure. To prevent COHb levels exceeding 2.5–3 per cent in non-smoking populations, the following guidelines have been proposed by WHO: a maximum permitted exposure of 100 mg m^{-3} (90 ppm) for 15 minutes; 60 mg m^{-3} (50 ppm) for 30 minutes; 30 mg m^{-3} (25 ppm) for 1 hour; and 10 mg m^{-3} (9 ppm) for 8 hours (WHO, 1987d, 1995a).

2.6 Lead

In those countries which do not yet use unleaded petrol, the addition of alkyl Pb additives in motor fuels accounts for an estimated 80–90 per cent of Pb in ambient air. The degree of pollution from this source differs from country to country, depending on motor vehicle density and efficiency of efforts to reduce the Pb content of petrol (WHO, 1987e). About 1 per cent of the Pb in petrol is emitted unchanged as tetraalkyl Pb (organic Pb). There is, in addition, some evaporation of organic Pb from the engine and fuel tank. Concentrations of tetraalkyl Pb amounting to more than 10 per cent of the total Pb content of ambient air have been measured in the immediate vicinity of service stations (NSIEM, 1983).

Most of the Pb in ambient air occurs as fine particles (10 μm). For adults, the retention rate of airborne particulate matter ranges from 20–60 per cent. Young children inhale a proportionately higher daily air volume per unit measure (weight, body area) than adults (Barltrop, 1972). It was estimated that children have a lung deposition rate of Pb that can be up to 2.7 times higher than that of adults.

The proportion of Pb absorbed from the gastrointestinal tract is about 0–15 per cent (Rabinowitz *et al.*, 1980) in adults, whereas levels of 40–50 per cent have been reported in children (Ziegler *et al.*, 1978). Absorption is influenced by dietary intake; fasting and diets with low levels of Ca, vitamin D, Fe and Zn have been shown to increase Pb absorption (Mahaffey, 1990). The unexcreted fraction of absorbed Pb is distributed among three bodily components: blood, soft tissues and mineralised tissues (bones, teeth). About 95 per cent of the body-burden of Pb in adults is located in the bones, compared with about 70 per cent in children. Non-absorbed Pb is excreted in the faeces. Of the absorbed fraction, 50–60 per cent is removed by renal and biliary excretion. The concentration of Pb in "milk" (baby) teeth provides a useful long-term record of Pb exposure in growing children.

Organic Pb compounds (tetraalkyl Pb and its metabolites) are volatile and liposoluble and are mainly taken up via the respiratory tract. Absorption by the lung is rapid and practically complete. Absorption through the skin is also important. Tetraalkyl Pb is metabolised in the liver and other tissues to trialkyl Pb, which is the most toxic metabolite (NSIEM, 1983).

Animal and epidemiological studies have demonstrated that Pb exposure may act on different bodily systems, principally haeme biosynthesis, the

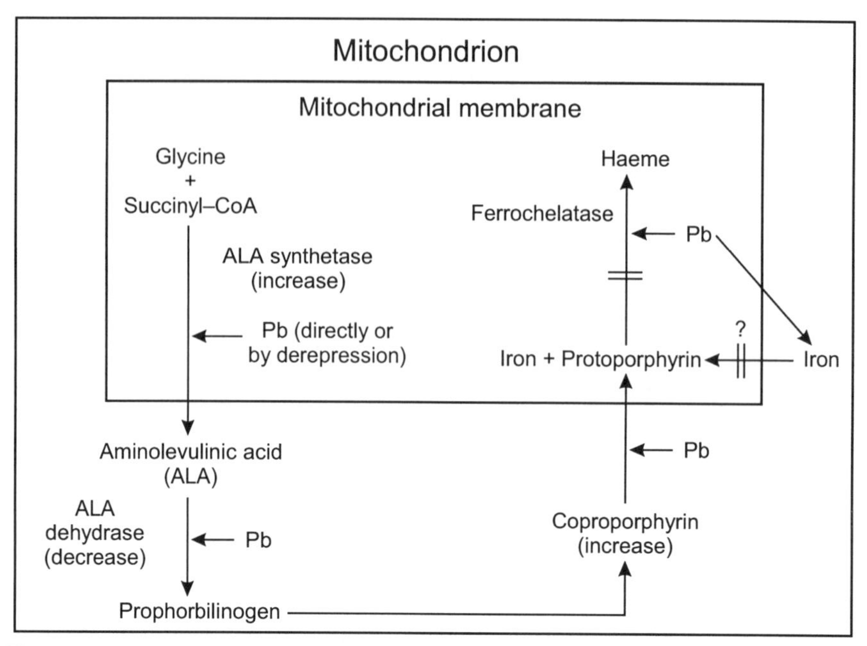

Figure 2.6 The effects of lead on haeme biosynthesis (After US EPA, 1986b)

nervous system, and others such as the cardiovascular system (blood pressure). Infants and young children less than five years old are particularly sensitive to Pb exposure because of its potential effect on neurological development. The effect of Pb on haeme biosynthesis and erythropoiesis occurs mainly through three enzymes:

- It stimulates the mitochondrial enzyme delta-aminolaevulinic acid synthetase (ALAS);
- It inhibits the activity of the cytoplasmic enzyme delta-aminolaevulinic acid dehydrase (ALAD) which results in an accumulation of its substrate ALA (this interference with haeme synthesis may occur at concentrations below 100 $\mu g\ l^{-1}$);
- It interferes with the functioning of intermitochondrial ferrochelatase, responsible for the insertion of Fe into the protoporphyrin ring, thereby resulting in an increase of erythrocyte protoporphyrin or zinc protoporphyrin in blood (Figure 2.6).

Anaemia is a frequent outcome of chronic Pb intoxication. Apart from this haematological effect, Pb also exerts an adverse effect on the endocrine system including the gonadal and reproductive systems (Rohn *et al.*, 1982). It also depresses thyroid function (Tuppurainen *et al.*, 1988) and impairs hepatic metabolism of cortisol (Saenger *et al.*, 1984). In young children, Pb exposure is also associated with a decrease in the biosynthesis of 1,25-dihydroxyvitamin D, an important metabolite of vitamin D (Mahaffey *et al.*, 1982).

The central nervous system is the primary target organ for Pb toxicity in children. Exposure to high concentrations of Pb can result in an encephalopathy which is more frequent in childhood Pb poisoning than in adult poisoning. The reason may be due to the ease with which Pb crosses the blood–brain barrier in children. Encephalopathy has occurred in children with blood Pb concentrations in excess of 800–1,000 $\mu g\ l^{-1}$ (NAS, 1972). The brain seems more sensitive to alkyl Pb exposure (NSIEM, 1983). Exposure of children to lower concentrations of Pb may produce neurophysiological disorders, including impairment of learning ability, behaviour, intelligence and fine motor co-ordination. Evidence for such disorders has been reviewed recently by (Davis and Svendsgaard, 1987; ATSDR, 1988; Grant and Davis, 1990). Needleman *et al.* (1979), in a community based study of children in Boston in whom previous exposure to Pb was estimated by examination of shed teeth, reported evidence of Pb-related neuropsychological deficits. This negative association between Pb concentration in teeth and mental development was also reported in subsequent studies (Winneke *et al.*, 1983, 1985). However, in other studies (Smith *et al.*, 1983; Harvey *et al.*, 1984), correcting for social environment greatly attenuated this inverse association.

More recently, some authors (Hawk *et al.*, 1986; Fulton *et al.*, 1987) have reported a significant inverse linear association between cognitive ability and blood Pb, with no evident threshold level of exposure. The mean blood Pb concentration of the group with the highest concentration in the study by Fulton *et al.* (1987) was 221 $\mu g\ l^{-1}$, suggesting that IQ (intelligence quotient) deficits are related to Pb exposure below 250 $\mu g\ l^{-1}$. In agreement with these findings, a recent study conducted in Mexico City among school children, from low to medium social status and aged 9–12 years, showed a strong negative correlation between blood Pb concentration and intellectual coefficients and teacher grading, without evidence of threshold concentration (Munoz *et al.*, 1993). Although none of the above studies can provide definite evidence that low-level Pb exposure is linked to reduced cognitive performance in children, the overall pattern of findings supports the conclusion that low-level Pb exposure is related to neurobehavioral dysfunction in children.

In a meta-analysis using data from 12 cross-sectional epidemiological studies, Needleman and Gatsonis (1990) concluded that blood Pb concentrations as low as 100–150 $\mu g\ l^{-1}$ are associated with IQ impairment in children.

Based on this body of data the lowest-observed adverse effect level has been defined as possibly < 100 μg l^{-1} (ATSDR, 1990).

Lead is transported to the foetus across the placenta because there is no metabolic barrier to foetal Pb uptake. Furthermore, the amount of Pb available for foetal uptake may actually be increased because part of the Pb stored in bone is released into the blood during pregnancy. Prenatal exposure to Pb produces toxic effects on the human foetus including reductions in gestational age, birth weight and mental development. These effects occur at relatively low blood Pb concentrations. An inverse association between maternal (or cord) blood Pb concentrations and gestational age has been reported by different authors (Dietrich *et al.*, 1986, 1987b; McMichael *et al.*, 1986). Based on risk estimates made by McMichael *et al.* (1986), the risk of premature delivery increases approximately four fold as cord or maternal blood Pb concentration increase from 80 μg l^{-1} to > 140 μg l^{-1}. Data from a Cincinnati study suggest an inverse relationship between prenatal maternal blood Pb concentration and birth weight and postnatal growth rates (Dietrich *et al.*, 1987a; Shulka, 1987). Other studies also support this inverse association (Bellinger *et al.*, 1984; Ward *et al.*, 1987).

A number of long-term studies currently in progress have investigated the effect of early Pb exposure and developmental effects. Bellinger *et al.* (1987, 1989) in Boston, studied the relationship between umbilical cord blood Pb and early cognitive development over 6 and 24 months of age. Lead concentrations were measured in 249 umbilical cord blood samples of infants born to middle and upper-middle class parents. Cord blood Pb concentrations were categorised as low (mean = 18 μg l^{-1}), mid (mean = 65 μg l^{-1}) and high (mean = 146 μg l^{-1}). After accounting for factors related to infant development, such as mother's age, race, IQ, education, care-giving environment, social class, and infant's sex, birth weight, birth order and gestational age, there was a significant inverse relation between performance on the Baylet Mental Development Index (MDI) at 6, 12 and 24 months and cord blood Pb concentrations (McMichael *et al.*, 1988).

The MDI is a composite scale to assess sensory-perceptual acuity, memory, learning ability, verbal communication and other cognitive functions (Shy, 1990). The difference in MDI between the low- and high-exposure groups was between 4 and 7 points. Postnatal blood Pb concentration showed no association with MDI deficit. Children in the lower socio-economic stratum were adversely affected at lower levels of prenatal exposure (Needleman, 1989). These findings are supported by other studies (Dietrich *et al.*, 1987a; Ernhart *et al.*, 1987).

Another major prospective study is being conducted in Port-Pirie, South Australia among a cohort of 537 children born during 1979 to 1982 to women living near a Pb smelter (McMichael *et al.*, 1988). Blood samples were collected from the mother, before and at delivery from the umbilical cord, and at ages 6, 15, 24 months and every year thereafter from the infants. At the age of two, the mean blood Pb concentration was 212 $\mu g\ l^{-1}$ with a range of 49–566 $\mu g\ l^{-1}$. The developmental status of each child was tested using the McCarthy Scale of Children's Abilities (MSCA). Maternal intelligence and care-giving environment were also evaluated. The blood Pb concentration at each age, particularly at 2 and 3 years and integrated postnatal average concentration were inversely related to development at the age of 4 years. Independently from other factors that may affect child development, subjects with an average postnatal blood Pb concentration of 309 $\mu g\ l^{-1}$ had a cognitive score seven points lower than those with an average concentration of 103 $\mu g\ l^{-1}$ (McMichael *et al.*, 1988).

A similar deficit occurred in the perceptual performance and memory score. There was no evidence that cognitive function at age 4 years was more influenced by recent than by earlier postnatal blood Pb concentrations. Within the range of exposure studied, there was no evidence of a threshold dose for an effect due to the presence of Pb. Further follow-up of this cohort has shown similar results among older children (McMichael *et al.*, 1994). The intelligence quotient declined by 2.6 points at age 7 (odds ratio of 2.6 with 90 per cent confidence interval between 0.13 and 4.9) for each natural-log unit increase in tooth Pb concentrations expressed in ppm. These results suggest that increased exposure to Pb results in developmental deficit, not just developmental delay.

Although the cognitive and neurosensory effects of low-level blood Pb are particularly difficult to study, mainly because of the variety of tests used and the number of different factors that may affect child development, there is an impressive convergence of animal and human studies (Shy, 1990). Grant and Davis (1990) concluded that neurobehavioral deficits and reduction in gestational age and birth weight are associated with prenatal internal exposure levels, indexed by maternal or cord blood Pb concentration of 100–150 $\mu g\ l^{-1}$ and possibly lower. In addition, a follow-up investigation of the children whose dentine concentrations of Pb had been measured in primary school showed that, 11 years later, children with high Pb level (dentine concentrations ≥ 20 ppm) were significantly more likely to drop out of high school and to have a reading disability (Needleman *et al.*, 1992). This study has been criticised because of potential bias (Good, 1991); nevertheless the results suggest that early Pb exposure may result in long-term neurobehavioural impairment.

In addition to the assessments discussed above, other aspects of Pb-associated neurotoxicity have also been examined. Hearing thresholds in children appear to be adversely affected by Pb. In the analysis of National Health and Nutrition Examination Survey (NHANES) II data by Schwartz and Otto (1987), the probability of Pb-induced hearing loss increased with increasing blood levels across the entire range of concentrations studied (between about 40 and 500 $\mu g\ l^{-1}$).

Exposure to high concentrations of Pb may lead to functional disorders of the gastrointestinal tract; a common sign of acute poisoning is colic. Lead may also produce damage in the kidneys, which leads to increased urinary excretion of amino acids, glucose and phosphate (Fanconi syndrome). After prolonged exposure, the injury may enter a chronic stage with fibrosis and arteriosclerotic changes in the kidney (Choie and Richter, 1980).

Epidemiological and animal data indicate that Pb increases blood pressure. In a study conducted in the USA, systolic and diastolic blood pressure were significantly related to blood Pb in white males aged 20–74 years, after adjusting for potential confounding variables (Pirkle *et al.*, 1985). These findings have been confirmed by another study (Pocock *et al.*, 1988). In a Danish follow-up study, Moller and Kristensen (1992) reported a decrease in blood pressure associated with a decrease in blood Pb concentrations over a five year period. However, the causal relationship between blood Pb concentrations and blood pressure is still unclear. The mechanism of toxicity could involve the Ca-mediated control of vascular smooth muscle contraction (Chai and Webb, 1988), and the renin-angiotensine system (Vander, 1988). The lowest-observed-effects levels for Pb-induced health effects in adults and children are shown in Tables 2.8 and 2.9.

In summary, the adverse effects of Pb exposure in early neurobehavioral development is of primary concern. It occurs at levels well below those considered "safe" in recent years. There can be little doubt that exposure to Pb, even at blood concentrations as low as 100–150 $\mu g\ l^{-1}$ and possibly lower, is linked to undesirable developmental outcomes in human foetuses and children (Davis and Svendsgaard, 1987). The most clearly identified effect has been lower scores on the MDI of the Baylet Scale of Infant Development, poor school attainment and lower intellectual coefficients, reduced gestational age, and lower birth weight. In different studies, a dose-response curve for blood Pb concentration and neurobehavioral impact has been derived (McMichael *et al.*, 1986, 1988; Bellinger *et al.*, 1987), and can be used to estimate the health impact of Pb exposure at the population level. In terms of implications for public health, an overall four point downward shift

Table 2.8 Summary of lowest observed effect levels (LOEL) for PbB in relation to key Pb-induced health effects in adults

	Health effects				
LOEL for PbB (μg/100 ml)	Haematological/ haeme synthesis	Neurological	Kidney function	Reproductive function	Cardiovascular
100–120		Encephalopathic signs and symptoms	Chronic nephropathy		
80	Frank anaemia				
60				Female reproductive effects	
50	Reduced haemoglobin projection	Overt subencephalopathic neurological symptoms		Altered testicular function	
40	Increased urinary ALA and elevated coproporphyrins	Peripheral nerve dysfunction (slowed nerve conduction)			
30					Elevated blood pressure (white males, aged 40–59)
25–30	Erythrocyte protoporphyrin (EP) elevation in males				
15–20	Erythrocyte protoporphyrin (EP) elevation in females				
< 10	ALA–D inhibition				

ALA Aminolevulinic acid

Source: ATSDR, 1990

in a normal distribution of Baylet MDI scores would result in 50 per cent more children scoring below 80 in this test.

It is also important to mention that lead absorption may be modulated by nutritional status. Nutrient deficiencies such as Ca, Fe and Zn have been associated with high blood Pb concentrations (Mahaffey and Michelson, 1980; Mahaffey, 1990) and, in some studies, Ca intake was inversely related

Table 2.9 Summary of lowest observed effect levels (LOEL) for PbB in relation to key Pb-induced health effects in children

	Health effects			
LOEL for PbB (μg/100 ml)	Haematological/ haeme synthesis	Neurological and related	Renal system	Gastrointestinal
80–100		Encephalopathic signs and symptoms	Chronic nephropathy (aminoaciduria, etc.)	Colic and other overt gastrointestinal symptoms
70	Frank anaemia			
60		Peripheral neuropathies		
50		?		
40	Reduced haemoglobin synthesis. Elevated coproporphyin. Increased urinary ALA	Peripheral nerve dysfunction. CNS cognitive effects (IQ deficits, etc.)		
30		?	Vitamin D metabolism interference	
15	Erythrocyte protoporphyrin elevation	Altered CNS electrophysiological responses		
10	ALA-D inhibition	MDI deficits, reduced gestational age and birth weight (prenatal exposure)	?	
	Py-5-N activity inhibition	?		
	?			

ALA Aminolevulinic acid
MDI Mental Development Index
CNS Central nervous system
Py-5-N Pyrimidine-5'-nucleotidase
Source: ASTDR, 1990

to blood Pb concentrations (Mahaffey, 1990; Sanin, 1995; Hernandez-Avil, 1996). Some epidemiological studies in Pb exposed populations are in progress to determine the impact of Ca supplementation on Pb absorption and mobilisation.

The use of Pb in petrol has been declining in various countries and this has been responsible for a substantial decrease in blood Pb concentrations in the general population (Annest *et al.*, 1983). A similar impact has been observed in Mexico among residents of Mexico City (Hernandez-Avila, 1995). Many countries, especially in the developing world, are still using leaded fuel. Considering that, in these countries, poor nutritional status and "home environment" may potentiate the effect of Pb exposure in a large proportion of the children, there is an urgent need for regulation and for continued research to identify other potential sources of Pb exposure and interacting factors with adverse effects, and to evaluate interventions.

The WHO guideline value for long-term exposure (e.g. annual average) to Pb in the air is 0.5 $\mu g\ m^{-3}$ (WHO, 1995b).

2.7 Benzene

Benzene is a constituent of crude oil. In Europe it is present in petrol in a proportion of about 5 per cent, occasionally as high as 16 per cent, although in the USA the benzene in petrol does not exceed 1.5–2 per cent. The major source of atmospheric benzene is emissions from motor vehicles and evaporation losses during handling, distribution and storage of petrol (WHO, 1987f). The benzene concentration of ambient air in residential areas is generally in the range 3–30 $\mu g\ m^{-3}$ (0.001–0.01 ppm) depending on vehicle traffic. Human benzene intake from the air may, therefore, range from 30 to 300 µg per day. The daily intake from food and water has been estimated to be 100–250 µg. People smoking 20 cigarettes per day would have a daily increased intake of approximately 600 µg (WHO, 1987f). About 50 per cent of inhaled benzene in the air is absorbed. Due to its high liposolubility, benzene is distributed mainly to the fat-rich tissues such as adipose tissue and bone marrow. Benzene is oxidised by the P-450-dependent oxidase system. Part of the absorbed benzene is exhaled unchanged in breath and part is eliminated in the urine after transformation.

The toxic effects of benzene in humans following inhalation exposure have been recently reviewed by WHO (1993) and include haematotoxicity, immunotoxicity, neurotoxicity and carcinogenicity. Three types of bone marrow effects have been reported in response to benzene exposure; these are bone marrow depression leading to aplastic anaemia, chromosomal changes, and carcinogenicity. Toxic effects have been observed with exposure to very high concentrations (more than 3,200 $mg\ m^{-3}$ or 1,000 ppm) with the appearance of neurotoxic syndrome. Acute poisoning can lead to death but high levels of exposure are associated with inflammation of the respiratory tract and haemorrhage of the lung. Persistent exposure to toxic levels may cause injury to the bone marrow, resulting in pancytopenia. This has been observed in several occupational studies in which workers were exposed to high benzene concentrations.

Benzene is a known human carcinogen classified by the International Agency on Research of Cancer (IARC), as Group 1 (IARC, 1982). Compounds classified as Group 1 are carcinogenic to humans. This category is used only when there is sufficient evidence of carcinogenicity in humans, i.e. a positive relationship has been observed between exposure to the agent and cancer and in which chance, bias and confounding could be ruled out with reasonable confidence.

A carcinogenic effect has been reported in workers exposed to benzene. These workers are more likely to develop acute leukaemia than the general population. Assessment of the risks of exposure to benzene has recently been reviewed using mathematical methods of extrapolation from high to low exposure (Van Raatle and Grasso, 1982). Using epidemiological data, the different mathematical models give estimates of excess leukaemia deaths ranging from 3 to 46 per thousand resulting from 30 years occupational exposure to 1 ppm benzene (IPCS, 1990). Estimated risks at lower exposure levels range from 0.08 to 10 excess leukaemia deaths per million population resulting from lifetime exposure of 1 $\mu g\ m^{-3}$ (approximately 0.0003 ppm) of benzene (Bailer and Hoel, 1989).

The Carcinogen Assessment Group (US EPA, 1985) estimated, using different mathematical models, that the "best-judgement" unit risk was 8.1×10^{-6}. These data can be used to assess human risk at low concentrations as seen in non-industrial communities. Thus, for example in the Los Angeles Basin where the population weighted concentration to benzene is 14.7 $\mu g\ m^{-3}$ (4.6 ppb), the added lifetime risk is estimated to be 101 to 780 cases per million people exposed, (SCAR, 1984). However, this method of estimating risk is not universally accepted because the mutagenic metabolite of benzene has not been identified, and because DNA repair systems may be more efficient at lower exposure concentrations (Read and Green, 1990). A study by Yin *et al.* (1989) reported significantly increased incidents of lung cancer as well as increased incidents of acute myelogenous leukaemia. Their results suggest that benzene might be a multi-site carcinogen in humans, as has been indicated in animal studies.

An expert task group of WHO has estimated the unit risk at 1 $\mu g\ m^{-3}$ of benzene to range between 4.4×10^{-6} and 7.5×10^{-6} (WHO, 1996b). There is no safe level for airborne benzene because benzene is carcinogenic for humans.

2.8 Polycyclic aromatic hydrocarbons

Polycyclic aromatic hydrocarbons (PAHs) are a group of chemicals formed during the incomplete combustion of wood and fuel. Exhaust from diesel engines contains lower concentrations of some gaseous pollutants but higher concentrations of particulate-bearing organic extracts, including PAHs. Other main sources of PAHs are coke production, heating with coal and cigarette smoke (high proportion). There are several hundred types of PAHs; the best-known is benzo[a]pyrene (BaP). Polycyclic aromatics are absorbed in the lung and gut. They are metabolised via the mixed-function oxidase system and the metabolites are thought to be the ultimate carcinogens (WHO, 1987g).

Evidence from experimental studies shows that many of these PAHs are mutagenic and carcinogenic. Epidemiological studies in coke oven workers and coal gas workers have suggested an increased risk of lung cancer in relation to PAH exposure (Steenland, 1986). More recently a case-control study of deaths among railway workers in the USA showed that workers aged under 65 years who had been exposed for at least 20 years, had a small but significantly increased risk of lung cancer (Garshick *et al.*, 1987). However, two recent studies among workers exposed to automotive emissions indicated that males usually employed as truck drivers or delivery men had a statistically significant (50 per cent) increase in the risk of bladder cancer (Silverman *et al.*, 1983, 1986). Based on studies using BaP as an index compound, the upper-bound estimate of lifetime cancer risk will be 62 per 100,000 exposed people per microgram of benzene-soluble coke oven emission per cubic metre of ambient air. Assuming a 0.71 per cent content of BaP in these emissions, it can be estimated that 9 out of 100,000 people exposed to 1 ng BaP m^{-3} over a lifetime would be at risk of developing cancer. There is no safe level for PAHs due to their carcinogenicity, and there is no known cancer threshold for BaP (WHO, 1987g). Benzo[a]pyrene has been classified as a human carcinogen and the unit risk has been estimated to be at 8.7×10^{-5} per ng m^{-3} (WHO, 1996b).

2.9 Aldehydes

Aldehydes are absorbed in the respiratory and gastrointestinal tract and metabolised. Most metabolites are excreted quickly, including bound formaldehyde. Acute irritant effects of aldehydes on human volunteers have been documented. For formaldehyde, these effects include ocular and olfactory irritation (observed at 0.06 mg m^{-3}), irritation of mucous membranes and alteration in respiration (observed at 0.12 mg m^{-3}), coughing, nausea and dyspnea (WHO, 1989). Allergic responses such as asthma and dermic allergy have also been observed.

Formaldehyde exposure has been associated with cancer risk mostly in occupational settings. The cancers most frequently encountered are nasal and nasopharyngeal, brain and leukaemia (Walrath and Fraumeni, 1983, 1984; Harrington and Oakes, 1984; Walrath and Fraumeni, 1984; Olsen and

Asnaes, 1986; Stroup *et al.*, 1986; Vaughan *et al.*, 1986a,b). Excess cancers in other body sites have also been described among occupationally exposed individuals. Human exposure to formaldehyde should be minimised, not only for its probable carcinogenic effect, but also for its potential for tissue damage. Epidemiological studies on carcinogenicity and that contain some exposure assessment imply that the threshold for tissue damage is between 0.5 and 3 mg m^{-3} (WHO, 1989). However, no risk estimate of carcinogenicity can be made because of lack of adequate data.

In summary, animal and epidemiological data conducted in occupational settings have been used to construct linear models to assess human risk of cancer at low concentrations seen in non-industrial communities. These models are available for benzene and BaP exposure but not for aldehydes. The carcinogenic risk to humans from diesel and petrol emissions has been evaluated by IARC, including all components in different experiments (IARC, 1989). The Agency concludes that diesel exhaust may be associated with lung and bladder cancer and it has been classified as probably carcinogenic to humans. Formaldehyde has also been classified as probably carcinogenic to humans (IARC 1987). There is no evidence of an association between a particular type of cancer and petrol exhaust and therefore it has been classified as possibly carcinogenic to humans.

In order to avoid irritation, the WHO guideline value for formaldehyde is fixed at 0.1 mg m^{-3} (0.083 ppm) as a 30 minute average. In the case of especially sensitive groups that show a hypersensitivity reaction without immunological signs, formaldehyde concentration should not exceed 0.01 mg m^{-3} (WHO, 1996b).

2.10 Conclusions

Epidemiological studies have been widely used to shed light on the effects of air pollutants on health due to vehicular traffic. In order to evaluate risks due to automotive emissions on the general population, several factors have to be considered: exposure, dose, biological effects, the dose-response relationship, and the proportion of population exposed. The results of some studies are difficult to interpret because of a variety of limitations, mostly regarding exposure assessment and handling of cofactors. Other studies provide enough information to derive dose-response functions that can be applied to ambient levels of specific pollutants in order to estimate selected health effects. These studies have been referred to in the text. For pollutants produced by vehicular emission, such estimates can be made to assess the potential adverse health effects of O_3, SPM, CO and Pb, as well as for the carcinogenic risk of exposure to benzene and BaP or PAHs. Various limitations of these estimates should be mentioned:

- Dose-response functions are population specific, and therefore their use in other populations may not be justified, especially if they have been derived from only one study.
- They do not consider the potential interactive effect of different pollutants.
- They are only mathematical models and are dependent on different sets of assumptions.
- Extrapolation of the shape of the curve outside of the range of observed values may lead to erroneous results e.g. it will be difficult to extrapolate to low level effects if there is no knowledge of the presence of a threshold value).
- In some cases, it is difficult to determine which is the most relevant exposure measurement for the health effect being studied.

However, despite their limitations, these mathematical models have great advantages:

- They allow a quantitative evaluation of the health impact of pollutants emitted by vehicles.
- They draw the attention of public health officials and the general public to the extent of the problem.
- They can be used for cost assessments.
- Cost effectiveness analyses can then be used to evaluate alternative control strategies.

Further research is needed in order to develop models more adapted to specific situations, and to develop techniques of biological monitoring (biomarkers) as indicators of exposure and early effects among the population.

2.11 References

Abbey, D.E., Peterson, F., Mills, P.K. and Beeson, W.L. 1993 Long-term ambient concentrations of total suspended particulates, ozone, and sulphur dioxide and respiratory symptoms in a non-smoking population. *Archives of Environmental Health,* **48**, 33–46.

Adler J.M. and Carey P.M. 1989 *Air toxics emissions and health risks from mobile sources*. 82nd Annual Meeting of the Air and Waste Association, June 1989, Anaheim, California.

Agostoni, A., Stabilini, R., Viggiano, G., Luzzana, M. and Samada M. 1980 Influence of capillary and tissue PO_2 on carbon monoxide binding to myoglobin: A theoretical evaluation. *Microvascular Research*, **20**, 81–87.

Alderman, B.W., Baron, A.E. and Savitz D.A. 1987 Maternal exposure to neighborhood carbon monoxide and risk of low birthweight. *Public Health Report*, **102**, 410–414.

Allred, E.N. Bleecker, E.R., Chaitman, B.R., Dahms, T.E., Gottlieb, S.O., Hackney, J.D., Pagano, M., Selvester, R.H., Walden, S.M. and Warren J. 1989 Short-term effects of carbon monoxide exposure on the exercise performance of subjects with coronary artery disease. *New England Journal of Medicine*, **321**, 426–432.

Annest, J.L., Pirkle, J.L., Neese, J.W., Bayse, D.D. and Kovar, M.G. 1983 Chronological trend in blood lead levels between 1976 and 1980. *New England Journal of Medicine*, **308**, 1373–1377.

Aris, R.M., Christian, D., Hearne, P.Q., Kerry, K., Finkbeiner, W.E. and Balmes, J.R. 1993 Ozone-induced airway inflammation in human subjects as determined by airway lavage and biopsy. *American Review of Respiratory Disease*, **148**, 1363–1372.

Aronow, W.S., Stemmer, E.A. and Isbell, M.W. 1974 Effect of carbon monoxide exposure on intermittent claudication. *Circulation*, **49**, 415–417.

Ash, S., Fisher, C.C., Truswell, A.S., Allen, J.R. and Irwig, L. 1989 Maternal weight gain, smoking and other factors in pregnancy as predictors of infant birth-weight in Sydney women. *Australian and New Zealand Journal of Obstetrics and Gynaecology*, **29**, 212–219.

ATSDR 1988 *The Nature and Extent of Lead Poisoning in Children in the United States: A Report to Congress.* Agency for Toxic Substances and Disease Registry, U.S. Department of Health and Human Services.

ATSDR 1990 *Toxicological Profile for Lead.* Agency for Toxic Substances and Disease Registry, U.S. Department of Health and Human Services.

Avol, E.L., Linn, W.S., Venet, T.G., Shamoo, D.A. and Hackney, J.D. 1984 Comparative respiratory effects of ozone and ambient oxidant pollution exposure during heavy exercise. *Journal of the Air Pollution Control Association*, **31**, 666–668.

Avol, E.L., Linn, W.S., Shamoo, D.A., Valencia, L.M., Anzar, U.T., Venet, T.G. and Hackney, J.D. 1985 Respiratory effects of photochemical oxidant air pollution in exercising adolescents. *American Review of Respiratory Disease*, **132**(3), 619–622.

Avol, E.L., Linn, W.S., Shamoo, D.A., Spier, C.E., Valencia, L.M., Venet, T.G., Trim, S.C. and Hackney, J.D. 1987 Short-term respiratory effects of photochemical oxidant exposure in exercising children, *Journal of the Air Pollution Control Association*, **37**, 158–162.

Avol, E.L., Navidi, W.C., Rapparport, E.B. and Peters, J.M. 1998 *Acute Effects of Ambient Ozone on Asthmatic, Wheezy and Healthy Children.* Special Report 82, Health Effects Insitute, Cambridge, MA.

Bailer, A.J. and Hoel, D.G. 1989 Benzene risk assessment: Review and update. *Cell Biology and Toxicology*, **5**, 287–295.

Barltrop, D. 1972 Children and environmental lead. In: P. Hepple [Ed.] *Lead in the Environment: Proceedings of a Conference*. Institute of Petroleum, London, 52–60.

Bascom, R., Bromberg, P.A., Costa, D.A., Devlin, R., Dockery, D.W., Frampton, M.W., Lambert, W., Samet, J.M, Speizer, F.E. and Utell M.J. 1996 (A committee of the environmental and occupational health assembly of the American Thoracic Society), 1995 State of the art review: Health effects of outdoor air pollution. *American Journal of Respiratory and Critical Care Medicine*, **153**, 3–150.

Bates, D.V. and Sizto, R. 1983 Relationship between air pollutant levels and hospital admissions in Southern Ontario, *Canadian Journal of Public Health*, **74**, 117–122.

Bates, D.V. and Sizto, R. 1987 Air pollution and hospital admissions in southern Ontario: The acid summer haze effect. *Environmental Research*, **43**, 317–331.

Beard, R.R. and Wertheim, G.A. 1967 Behavioral impairment associated with small doses of carbon monoxide. *American Journal of Public Health*, **57**, 2012–2022.

Bellinger, D.C., Needleman, H.L., Leviton, A., Waternaux, C., Rabinowitz, M.B. and Nichols, M.L. 1984 Early sensory-motor development and prenatal exposure to lead. *Neurobehavioral Toxicology and Teratology*, **6**, 387–402.

Bellinger, D., Leviton, A., Waternaux, C. and Rabinowitz, M. 1987 Longitudinal analyses of prenatal and postnatal lead exposure and early cognitive development. *New England Journal of Medicine*, **316**, 1037–1043.

Bellinger, D., Leviton, A., Waternaux, C., Needleman, H. and Rabinowitz M. 1989 Low level lead exposure, social class, and infant development. *Neurotoxicology and Teratology*, **10**, 497–503.

Borja-Aburto, V., Loomis, D.P., Shy, C. and Bangdiwala, S. 1995 Air pollution and daily mortality in Mexico City. *Epidemiology*, **4**, S64.

Braun-Fahrlander, C., Ackermann-Liebrich, U., Schwartz, J., Gnehm, H.P., Rutishauser M. and Wanner, H.U. 1992 Air pollution and respiratory symptoms in preschool children. *American Review of Respiratory Health*, **145**, 42–47.

Bumgartner, J.R. and Speizer, F. 1991 Chronic obstructive pulmonary disease. Unpublished paper, Population, Health and Nutrition Division, World Bank, Washington D.C.

Burnett, R.T., Dales, R.E., Raizenne, M.E., Krewski, D., Summers, P.W., Roberts, G.R., Raad-Young, M., Dann, T. and Brook, J. 1994 Effects of low ambient levels of ozone and sulfates on the frequency of respiratory admissions to Ontario Hospitals. *Environmental Research*, **65**,172–194.

Calabrese, E.J., Moore, G.S., Guisti, R.A., Rowan, C.A. and Schultz, E.N. 1981 A review of human health effects associated with exposure to diesel fuel exhaust. *Environment International*, **5**, 473–477.

Campbell, M.J., Lewry, J. and Wailoo, M. 1988 Further evidence for the effect of passive smoking on neonates. *Postgraduate Medical Journal*, **64**, 663–665.

Castillejos, M., Gold, D., Dockery, D., Tosteson, T., Baum, T. and Speizer, F. 1992 Effects of ambient ozone on respiratory function and symptoms in school children in Mexico City. *American Review of Respiratory Health*, **145**, 276–282.

Chai, S. and Webb, R.C. 1988 Effects of lead on vascular reactivity. *Environmental Health Perspectives*, **78**, 53–56.

Choie, D.D. and Richter, G.W. 1980 Effects of lead in the kidney. In: R.L. Singhal, and J.A Thomas, [Eds] *Lead Toxicity*. Urban and Schwarzenberg, Baltimore, 337–350.

Cody, R.P.;, Weisel, C.P., Birnbaum, G. and Lioy, P.J. 1992 The effect of ozone associated with summertime photochemical smog on the frequency of asthma visits to hospital emergency departments. *Environmental Research*, **58**, 184–194.

Damji, K.S. and Richters, A. 1989 Reduction of T lymphocyte subpopulations following acute exposure to 4 ppm nitrogen dioxide. *Environmental Research*, **49**, 217–224.

Davis, M.J. and Svendsgaard, D.J. 1987 Lead and child development. *Nature*, **329**, 297–300.

Detels, R., Tashkin, D.P., Sayre, J.W., Rokaw, S.N., Coulson, A.H., Massey, F.J. Jr. and Wegman, D.H. 1987 The UCLA population studies of chronic obstructive respiratory disease. 9. Lung function changes associated with chronic exposure to photochemical oxidants, A cohort study among never smokers. *Chest*, **92**, 594–603.

Devlin, R.B., McDonnell, W.F., Mann, R., Becker, S., House, D.E., Schreinemachers, D. and Koren, H.S. 1991 Exposure of humans to ambient levels of ozone for 6.6 hours causes cellular and biochemical changes in the lung. *American Journal of Respiratory Cell and Molecular Biology*, **4**,72–81.

Devlin, R., Horstman, D., Becker, S., Gerrity, T., Madden, M. and Koren, H. 1992 Inflammatory response in humans exposed to 2.0 ppm NO. *American Review of Respiratory Disease*, **145**, A455.

Dietrich, K.N., Krafft, K.M. and Bier, M. 1986 Early effects of lead exposure: neurobehavioral findings at 6 months. *International Journal of Biosocial Research*, **8**, 151–168.

Dietrich, K.N., Krafft, K.M., Bier, M., Berger, O., Succop, P.A. and Bornschein, R.L. 1990 Neurobehavioral effects of fetal lead exposure: the first year of life. In: M. Smith, L. D. Grant and A. I. Sors [Eds] *Lead Exposure and Child Development: An International Assessment*. MTP Press, Lancaster, UK.

Dietrich, K.N., Krafft, K.M., Bornschein, R.L., Hammond, P.B., Berger, O., Succop, P.A. and Bier, M. 1987b Low-level fetal lead exposure effect on neurobehavioural development in early infancy. *Pediatrics*, **80**(5), 721–730.

Dietrich, K.N., Krafft, K.M., Shukla, R., Bornschein, R.L. and Succop, P.A. 1987a The neurobehavioral effects of early exposure. In: S. R. Schroeder, [Ed] *Toxic Substances and Mental Retardation: Neurobehavioral Toxicology and Teratology*. Monographs of the American Association on Mental Deficiency: No. 8., M. J. Begab [Ed], American Association on Mental Deficiency, Washington D.C., 71–95.

Dijkstra, L., Houthuijs, D., Brunekreef, B., Akkermann, I. and Boleij, J.S.M. 1990 Respiratory health effects of the indoor environment in a population of Dutch children. *American Review of Respiratory Disease*, **142**,1172–1178.

Dockery, D.W., Ware, J.H., Ferris, G. Jr., Speizer, F.E and Cook, N.R. 1982 Change in pulmonary functions in children associated with air pollution episodes. *Journal of the Air Pollution Control Association*, **23**, 937–942.

Dockery, D.W., Speizer, F.E., Stram, D.O., Ware, J.H., Spengler, J.D. and Ferris, B. G. Jr. 1989 Effects of inhalable particles on respiratory health of children. *American Review of Respiratory Disease*, **139**, 587–594.

Dockery, D.W. and Pope, C.A. 1994 Acute respiratory effects of particulate air pollution. *Annual Review of Public Health*, **15**, 107–132.

Elsayed, N.M., Kass, R., Mustafa, M.G., Hacker, A.D., Ospital, J.J., Chow, C.K. and Cross, C.E. 1988 Effect of dietary vitamin E Level on the biochemical response of rat lung to ozone inhalation. *Drug Nutrient Interactions*, **5**, 373–386.

Emmelin, A., Nystrom, L. and Wal, S. 1993 Diesel exhaust exposure and smmoking: a case reference study of lung cancer among Swedish workers. *Epidemiology*, **4**(3), 237–244.

Ernhart, C.B., Morrow-Tlucak, M., Marler, M.R. and Wolf, A.W. 1987 Low level lead exposure in the prenatal and early preschool periods: Early pre-school development. *Neurotoxicology and Teratology*, **9**, 259–270.

Euler, G.L., Abbey, D.E., Magie, A.R. and Kodkin, J.E. 1987 Chronic obstructive pulmonary disease symptom effects of long term cumulative exposure to ambient levels of total suspended particulates and sulfur dioxide in California Seventh-Day Adventist residents. *Archives of Environmental Health*, **42**, 213–222.

Euler, G.L., Abbey, D.E., Hodgkin, J.E. and Magie, A.R. 1988 Chronic obstructive pulmonary disease symptom effects of long-term cumulative exposure to ambient levels of total oxidants and nitrogen dioxide in California Seventh-day Adventist Residents. *Archives of Environmental Health*, **43**, 279–285.

Evans, J.S., Tosteson, T. and Kinney, P.L. 1984 Cross-sectional mortality studies and air pollution risk assessment. *Environment International*, **10**, 55–83.

Folinsbee, L.J. 1989 Human health effects of exposure to airborne acid. *Environmental Health Perspectives*, **79**, 195–199.

Folinsbee, L.J., Hortman, D.H., Kehrt, H.R., Harder, S., Abdul-Salaan, S. and Ives, P.J. 1994 Respiratory responses to repeated prolonged exposure to 0.12 ppm ozone., *American Journal of Respiratory and Critical Care Medicine*, **149**, 98–105.

Folinsbee, L.J., McDonnell, W.F. and Horstman, D.H. 1988 Pulmonary function and symptom responses after 6.6-hour exposure to 0.12 pm ozone with moderate exercise. *Journal of the American Pollution Control Association*, **38**(1), 28–35.

Fulton, M., Raab, G., Thomson, G., Laxen, D., Hunter, R. and Hepburn, W. 1987 Influence of blood lead on the ability and attainment of children in Edinburgh. *Lancet*, **1** (8544), 1221–1226.

Garshick, E., Schenker, M.B., Munoz, A., Segal, M., Smith, T.J., Woskie, S.R., Hammond, S.K. and Speizer, F.E. 1987 A case control study of lung cancer and diesel exhaust exposure in railroad workers. *American Review of Respiratory Disease*, **135**, 1242–1248.

Garshick, E., Schenker, M.B., Munoz, A., Segal, M., Smith, T.J., Woskie, S.R., Hammond, S.K. and Speizer, F.E. 1988 A retrospective cohort study of lung cancer and diesel exhaust exposure in railroad workers. *American Review of Respiratory Disease*, **137**(4), 820–825.

Gibbons, S.I. and Adams, W.C. 1984 Combined effects of ozone exposure and ambient heat on exercising females. *Journal of Applied Physiology*, **57**(2), 450–456.

Good, M.I. 1991 The long-term effects of exposure to low doses of lead in childhood. *New England Journal of Medicine*, **324**, 415–418.

Graham, D.E. and Koren, H.S. 1990 Biomarkers of inflammation in ozone-exposed humans. *American Review of Respiratory Disease*, **142**, 152–156.

Grant, L.D. and Davis, J.M. 1990 Effects of low-level lead exposure on pediatric neurobehavioral and physical development: Current findings and future directions. In: M. Smith, L.D. Grant and A. Sors. [Eds] *Lead Exposure and Child Development: An International Assessment.* MTP Press, Lancaster, UK.

Gustavsson, P., Plato, N., Lidstrom, E.B. and Hogstedt, C. 1990 Lung cancer and exposure to diesel exhaust among bus garage workers. *Scand. J. Work. Environ. Health*, **16**(5), 348–354.

Hackney, J.D., Linn, W.S. and Avol, E.L. 1989 Acid fog: effects on respiratory function and symptoms in healthy asthmatic volunteers. *Environmental Health Perspectives*, **79**, 159–162.

Harrington, J.M. and Oakes, D. 1984 Mortality study of British pathologists, 1974–1980. *British Journal of Industrial Medicine*, **41**, 188–191.

Harvey, P.G., Hamlin, M.W., Kumar, R. and Delves, I.T. 1984 Blood lead, behavior and intelligence test performance in pre-school children. *Science of the Total Environment*, **40**, 45–60.

Hasselblad, V., Eddy, D.M. and Kotchmar, D.J. 1992 Synthesis of environmental evidence: nitrogen dioxide epidemiology studies. *Journal of the Air Waste Management Association*, **42**, 662–671.

Hawk, B.A., Schroeder, S.R., Robinson, G., Otto, D., Mushak, P., Kleinbaum, D. and Dawson, G. 1986 Relation of lead and social factors to IQ of low-SES children: A partial replication. *American Journal of Mental Deficiency*, **91**, 178–183.

Hazucha, M.J., Seal, E., Folinsbee, L.J. and Bromberg, P.A. 1994 Lung function response of healthy women after sequential exposures to NO_2 and O_3. *American Journal of Respiratory and Critical Care Medicine*, **150**, 642–647.

HEI (Health Effect Institute) 1988 *Air Pollution, the Automobile and Public Health.* National Academic Press, Washington D.C.

Hebel, J.R., Fox, N.L. and Sexton, M. 1988 Dose-response of birth weight to various measures of maternal smoking during pregnancy. *Journal of Clinical Epidemiology*, **41**, 483–489.

Hernandez-Avila, M. 1995 El plomo: un problema de salud publica en Mexico, In: M. Hernandez-Avila, and R.E Palazuelos [Eds] Intoxicacion por plomo en Mexico, prevencion y control, *Perspectiva en Salud Publica*, **21**, Instituto Nacional de Salud Publica, Guernavaca, Mexico, 13–24.

Hernandez-Avila, M., Gonzales-Cossio, T., Palazuelos, E., Romieu, I., Aro, A., Fishbein, E., Peterson, K. and Hu, H. 1996 Dietary and environmental determinants of blood and bone lead in lactating post-partum women living in Mexico City. *Environmental Health Perspectives*, **104**(10), 1076–1082.

Hoek, G. and Brunekreef, B. 1993 Acute effects of a winter air pollution episode on pulmonary function and respiratory symptoms of children. *Archives of Environmental Health*, **48**, 328–335.

Hoek, G., Fischer, P., Brunekreef, B., Lebret, E., Hofschreuder, P. and Mennen, M.G. 1993 Acute effects of ambient ozone on pulmonary function on children in the Netherlands. *American Review of Respiratory Disease*, **147**, 111–117.

Holguin, A.H., Buffler, P.A., Contant, C.F., Stock, T.H., Kotchmar, D., Hsi, B.P., Jenkings, D.E., Gehan, B.M., Noel, L.M. and Mei, M. 1994 The effects of ozone on asthmatics in the Houston area. In: Si Duk Lee [Ed.] *Evaluation of the Scientific Basis for Ozone/Oxidants Standards* (transactions). Proceedings of the Air Pollution Control Association International Specialty Conference, Houston, Texas.

Horstman, D.H., Seal, E.J., Folinsbee, L.J., Ives, P. and Roger, L.H. 1988 The relationships between exposure duration and sulfur dioxide-induced bronchoconstruction in asthmatic subjects. *American Industrial Hygiene Association Journal*, **49**, 38–47.

Imai, M., Yoshida, K., Kotchmar, D.J. and Lee, S.D. 1985 A survey of health studies of photochemical air pollution in Japan. *Journal of the Air Pollution Control Association*, **35**, 103–108.

IARC 1982 Benzene and Annex, In: *Some Industrial Chemicals and Dyestuffs*. Monographs on the evaluation on carcinogenic risk of chemicals to humans, Volume **29**, International Agency for Research on Cancer, Lyon, France.

IARC 1983 *Polynuclear aromatic compounds, Part I, Chemical, Environmental and Experimental Data.* Monographs on the evaluation of carcinogenic risk of chemicals to humans, Volume **32,** International Agency for Research on Cancer. Lyon, France.

IARC 1987 *Chemicals, Industrial Processes and Industries Associated with Cancer in Humans*, Monographs on the evaluation on carcinogenic risk of chemicals to humans, supplement **7**, International Agency for Research on Cancer, Lyon, France.

IARC 1989 *Diesel and Gasoline Engine Exhaust and Some Nitroarenes*, Monographs on the evaluation on carcinogenic risk of chemicals to humans, Volume, **46**, International Agency for Research on Cancer. Lyon, France.

IPCS 1990 Working document on human health and environmental effects of motor vehicle fuels and exhaust emissions, PCS/WP/90.63 International Programme on Chemical Safety, Geneva.

Katsouyanni, K., Zmirou, D., Spix, C., Sunyer, J., Schouten, J.P., Ponka, A., Anderson, H.R., Le Moullec, Y., Wojtyniak, B., Vigotti, M.A. and Bacharova, L. 1995 The APHEA project: background, objectives, design. Short-term effects of air pollution on health, A European approach using epidemiological time series data. *European Respiratory Journal*, **8**, 1030–1038.

Kilburn, K.H., Warshaw, R.H. and Thornton, J.C., 1992 Expiratory flows decreased in Los Angeles children from 1984 to 1987: Is this evidence for effects of air pollution? *Environmental Research*, **59**, 150–158.

Kinney, P.L. and Ozkaynak, H., 1991 Associations of daily mortality and air pollution in Los Angeles County. *Environmental Research*, **54**, 99–120.

Kinney, P.L. and Ozkaynak, H. 1992 Association between ozone and daily mortality in Los Angeles and New York City. *American Review of Respiratory Disease*, **145**(4:2), A:95.

Kinney, P.L., Ware, J.H. and Spengler, J.D. 1988 A critical evaluation of acute ozone epidemiology results. *Archives of Environmental Health*, **43**(2), 168–173.

Kleinman, M.T., Lurman, F.W., Winer, A.M., Colome, S.D., Brajer, V. and Hall, J.V. 1989a *Effects on Human Health of Pollutants in the South Coast Air Basin*, Final Report for South Coast Air Quality Management District, California State University Fullerton Foundation.

Kleinman, M.T., Phalen, R.F., Mautz, W.J., Mannix, R.C., McLure, T.R. and Crocker, T.T. 1989b Health effect of acid aerosols formed in atmospheric mixtures. *Environmental Health Perspectives*, **79**, 137–145.

Koenig, J.Q., Covert, D.S., Marshall, S.G., Van Belle, G. and Pierson, WE. 1989a The effects of ozone and nitrogen dioxide on pulmonary function in healthy and asthmatic adolescents, *American Review of Respiratory Disease*, **136**, 1152–1157.

Koenig, J.Q., Covert, D.S., Hanley, Q.S., Van Belle, G. and Pierson, W.E. 1989b Prior exposure to ozone potentiates subsequent response to sulfur dioxide in adolescent asthmatic subjects. *American Review of Respiratory Disease*, **141**, 377–380.

Koren, H.S., Devlin, R.B., Becker, S., Perez, R. and McDonnell, W.F. 1991 Time-dependent changes of markers associated with inflammation in the lungs of humans exposed to ambient levels of ozone. *Toxicologic Pathology*, **19**, 406–411.

Krzyzanowski, M., Quackennboss, J.J. and Lebowitz, M.D. 1992 Relation of peak expiratory flow rates and symptoms to ambient ozone. *Archives of Environmental Health*, **47**, 102–115.

Kulle, T.J., Sauder, L.R., Hebel, J.R. and Chatham, M.D. 1985 Ozone response relationships in healthy nonsmokers. *American Review of Respiratory Disease*, **132**, 36–41.

Lambert, P.M. and Reid D.D. 1970 Smoking, air pollution and bronchitis in Britain, *Lancet*, **7636**, 853–857.

Lebowitz, M.D., Collins, L., Holberg, C.H. 1987 Time series analyses of respiratory responses to indoor and outdoor environmental phenomena. *Environment*, **43**, 332–341.

Linn, W.S., Avol, E.L., Peng, R.C., Shamoo, D.A. and Hackney, J.D. 1987 Replicated dose-response study of sulfur dioxide effects in normal, atopic and asthmatic volunteers. *American Review of Respiratory Disease*, **136**, 1127–1134.

Linn, W.S., Avol, E.L., Shamoo, D.A., Spier, C.E., Valencia, L.M., Venet, T.G., Fischer, D.E, and Hackney, J.D. 1986 A dose-response study of healthy, heavily exercising men exposed to ozone at concentrations near the ambient air quality standard. *Toxicology and Industrial Health*, **2**, 99–112.

Lioy, P.J., Vollmuth, T.A. and Lippmann, M. 1985 Persistence of peak flow decrement in children following ozone exposures exceeding the National Ambient Air Quality Standard. *Journal of the Air Pollution Control Association*, **35**(10), 1069–1071.

Lippmann, M. 1989a Effects of ozone on respiratory function and structure. *Annual Review of Public Health*, **10**, 49–67.

Lippmann, M. 1989b Health effects of ozone: A critical review. *Journal of the Air Pollution Control Association*, **39**, 672–695.

Lippmann, M. 1989c Background on health effects of acid sulfate aerosols. *Environmental Health Perspectives*, **79**, 3–6.

Lippmann, M. 1993 Health effects of tropospheric ozone. Review of recent research findings and their implications to ambient air quality standards. *Journal of Exposure Analysis and Environmental Epidemiology*, **3**, 103–129.

Lippmann, M., Lioy, P.J., Leihauf, G., Green, K.B. and Baxter, D. 1983 Effects of ozone on the pulmonary funciton of children. *Environmental Toxicology*, **5**, 423–446.

Logan, W.P.D. 1953 Mortality in London fog incident. *Lancet*, **265**, 336–338.

Loomis, D.P., Borja-Aburto, V.H., Bangdiwala, S.I. and Shy, C.M. 1996 *Ozone Exposure and Daily Mortality in Mexico City: A Time-Series Analysis*. Research Report Number 75, Health Effects Institute, Cambridge, MA.

Mahaffey, K.R. 1990 Environmental lead toxicity: Nutrition as a component of intervention. *Environmental Health Perspectives*, **89**, 75–78.

Mahaffey, K.R. and Michelson, I.A. 1980 The interaction between lead and nutrition. In: H.L. Needleman [Ed.] *The Clinical Implications in Current Research*, Raven Press.

Mahaffey, K.R., Annest, J.L., Roberts, J. and Murphy, R.S. 1982 National estimates of blood lead levels: United States, 1976–1980: Association with selected demographic and socioeconomic factors. *New England Journal of Medicine*, **307**, 573–579.

Mathai, M., Skinner, A., Lawton, K. and Weindling, A.M. 1990 Maternal smoking, urinary cotinine levels and birth-weight. *Australian and New Zealand Journal of Obstetrics and Gynaecology*, **30**, 33–36.

McDonnell, W.F., Chapman, R.S., Leigh, M.W., Strope, G.L. and Collier, A.M. 1985 Respiratory responses of vigorously exercising children to 0.12 ppm ozone exposure. *American Review of Respiratory Disease*, **132**(4), 875–879.

McMichael, A.J., Vimpani, G.V., Robertson, E.F., Baghurst, P.A. and Clark, P.D. 1986 The Port Pirie cohort study: maternal blood lead and pregnancy outcome. *Journal of Epidemiology and Community Health*, **40**, 18–25.

McMichael, A.J., Baghurst, P.A., Wigg, N.R., Vimpani, G.V., Robertson, E.F. and Roberts, R.J. 1988 Port Pirie cohort study: Environmental exposure to lead and children's abilities at the age of four years. *New England Journal of Medicine*, **319**, 468–475.

McMichael, A.J., Baghurst, P.A., Vimpani, G.V., Wigg, N.R., Robertson, E.F. and Tong, S. 1994 Tooth lead levels and IQ in school-age children: The Port Pirie cohort study. *American Journal of Epidemiology*, **140**, 489–499.

Messineo, T.D. and Adams, W.C. 1990 Ozone inhalation effects in females varying widely in lung size: comparison with males. *Journal of Applied Physiology,* **69**, 96–103.

Ministry of Health, 1954 *Mortality and Morbidity During the London Fog of December 1952*, Reports on Public Health and Medical Subjects No 95, Ministry of Health, London.

Mohsenin, V. 1987a Airway responses to nitrogen dioxide in asthmatic subjects. *Journal of Toxicology and Environmental Health*, **22**, 371–380.

Mohsenin V. 1987b Effects of vitamin C on NO_2 induced airway hyperresponsiveness in normal subjects: a randomized double-blind experiment. *American Review of Respiratory Disease*, **136**, 1408–1411.

Moller, L. and Kristensen, T.S. 1992 Blood lead as a cardiovascular risk factor. *American Journal of Epidemiology*, **136**, 1091–1100.

Monson, R.R. and Nakano, K.K. 1976 Mortality among rubber workers. I. White male union employees in Akron, Ohio. *American Journal of Epidemiology*, **103**, 284–296.

Morrow, P.E. and Utell, M.J. 1989 Responses of susceptible subpopulations to nitrogen dioxide. *Research Report, Health Effects Institute*, **23**, 1–44.

Morrow, P.W. 1984 Toxicological data on NO_x an overview. *Journal of Toxicology and Environmental Health*, **13**, 205–227.

Munoz, H., Romieu, I., Palazuelos, E., Mancilla-Sanchez, T. and Meneses-Gonzales, F. 1993 Blood lead level and neurobehavioral development among children in Mexico City. *Archives of Environmental Health*, **48**, 132–139.

NAS 1972 *Lead: Airborne Lead in Perspective.* Committee on biological effects of atmospheric pollutants, National Academy of Sciences, Washington, D.C.

NSIEM 1983 *Health Risks Resulting from Exposure to Motor Vehicle Exhaust.* A report to the Swedish Government, Committee on Automotive Air Pollution, National Swedish Institute of Environmental Medicine, Stockholm.

Neas, L.M., Dockery, D.W., Ware, J.H., Spengler, J.D., Speizer, F.E. and Ferris, B.G. Jr. 1991 Association of indoor nitrogen dioxide with respiratory symptoms and pulmonary function in children. *American Journal of Epidemiology*, **134**, 204–219.

Neas, L.M., Dockery, D.W., Spengler, J.D., Speizer, F.E. and Tollerud, D.J. 1992 The association of ambient air pollution with twice daily peak expiratory flow measurements in children. *American Review of Respiratory Disease*, **145**, A429.

Neas, L.M., Dockery, D.W., Koutrkis, P., Tollerud, D.J. and Speizer, F.E., 1995 The association of ambient air pollution with twice daily peak expiratory flow rate measurements in children. *American Journal of Epidemiology*, **141**, 111–122.

Needleman, H.L., Gunnoe, C., Leviton, A., Reed, R., Peresie, H., Maher, C. and Barrett, P. 1979 Deficits in psychological and classroom performance of children with elevated dentine lead levels. *New England Journal of Medicine*, **300**, 689–695.

Needleman, H.L. 1989 The persistent threat of lead: A singular opportunity. *American Journal of Public Health*, **75**, 643–645.

Needleman, H.L. and Gatsonis, C.A. 1990 Low-level lead exposure and the I.Q. of children, A meta-analysis of modern studies. *Journal of American Medical Association*, **263**, 673–678.

Needleman, H.L., Schell, A., Bellinger, D., Leviton, A. and Allred, E.N. 1992 The long-term effects of exposure to low doses of lead in childhood. An 11 year follow-up report. *New England Journal of Medicine*, **322**, 83–88.

Olsen, J.H. and Asnaes, S. 1986 Formaldehyde and the risk of sqamous cell carcinoma of the sinonasal cavities. *British Journal of Industrial Medicine*, **43**, 769–774.

Ostro, B.D. 1989 Estimating the risk of smoking, air pollution and passive smoke on acute respiratory conditions. *Risk Analysis*, **9**, 189–191.

Ostro, B.D. 1990 Transferring air pollution health effects across European borders: Issues of measurement and efficiency. Paper presented to the

International Conference on Environmental Co-operation and Policy in the Single European Market, Venice, Italy, April 17–20.

Ostro, B.D. 1994 Search for a threshold in the relation of air pollution to mortality: A reanalysis of data on London mortality. *Environmental Health Perspectives*, **58**, 397–399.

Ostro, B.D., Lipsett, M.J., Mann, J.K., Braxton-Owens, H. and White, M.C. 1995 Air pollution and asthma exacerbations among African-American children in Los Angelis, *Inhalation Toxicology* **7**, 711–722.

Ostro, B.D. Lipsett M.J., Wiener M.B. and Selner, J.C. 1991 Asthmatics' responses to airborne acid aerosols. *American Journal of Public Health*, **81**, 694–702.

Ostro, B.D., Sanchez, J.M., Aranda, C. and Eskeland, G. 1996 Air pollution and mortality: Results from a study of Santiago, Chile. *Journal of Exposure Assessment and Environmental Epidemiology*, **6**(1), 97–114.

Ott, M.G., Townsend, J.C., Fishbeck, W.A. and Langner, R.A., 1978 Mortality among individuals occupationally exposed to benzene. *Archives of Environmental Health*, **33**, 3–10.

Peters, A., Tuch, T., Heinrich, J., Heyder, J. and Wichmann, H.E. 1995 Short-term effects of PM_{10} fine and ultra-fine particles, on lung function and symptoms. *Epidemiology*, **6** (4), S64.

Pirkle, J.L., Schwartz, J., Landis, J.R. and Harlan, W.R. 1985 The relationship between blood lead levels and blood pressure and its cardiovascular risk implications, *American Journal of Epidemiology*, **121**, 246–258.

Pocock, S.J., Shaper, A.G., Ashby, D., Delves, H.T. and Clayton, B.E. 1988 The relationship between blood lead, blood pressure, stroke, and heart attacks in middle-aged British men. *Environmental Health Perspectives*, **78**, 23–30.

Pope, C.A., Bates, D.V. and Raizenne, M.E. 1995 Health effects of particulate air pollution: Time for reassessment? *Environmental Health Perspectives*, **103**, 472–480.

Pope, C.A., Dockery, D.W., Spengler, J.D. and Raizenne, M.E. 1991 Respiratory health and PM_{10} pollution: A daily time series analysis. *American Review of Respiratory Disease*, **144**, 668–674.

Rabinowitz, M.B., Kopple, J.D. and Wetherill, G.W. 1980 Effect of food intake and fasting on gastrointestinal lead absorption in humans. *American Journal of Clinical Nutrition*, **33**, 1784–1788.

Raizenne, M., Burnett, R.T., Stern, B., Franklin, C.A. and Spengler, J.D. 1989 Acute lung function response to ambient acid aerosol exposure in children. *Environmental Health Perspectives*, **79**, 179–185.

Read, R.C. and Green, M. 1990 Internal combustion and health. *British Medical Journal*, **300**, 761–762.

Roemer, W., Hoeck, G. and Brunekreef, 1993 Effects of ambient winter air pollution on respiratory health of children with chronic respiratory symptom. *American Review of Respiratory Disease*, **147**, 118–124.

Roger, L.J., Horstman, D.H., McDonnell, W., Kehrl, H., Ives, P.J., Seal, E., Chapman, R. and Massaro, E.J. 1990 Pulmonary function, airway responsiveness, and respiratory symptoms in asthmatics following exercise in NO_2. *Toxicology and Industrial Health*, **6**, 155–171.

Rohn, R.D., Shelton, J.E. and Hill, J.R. 1982 Somatomedin activity before and after chelation therapy in lead-intoxicated children. *Archives of Environmental Health*, **37**, 369–373.

Romieu, I. and Borja-Aburto, V.H. 1997 Particulate air pollution and daily mortality: can results be generalized to Latin American countries? *Salud Publica de Mexico*, **39**(5), 403–411.

Romieu, I., Cortes Lugo, M., Ruiz-Velasco, S., Sanchez, S., Meneses, F. and Hernandez, M. 1992 Air pollution and school absenteeism among children in Mexico City. *American Journal of Epidemiology*, **136**, 1524–1531.

Romieu, I., Meneses, F., Ruiz-Velasco, S., Sienra-Monge, J.J., Huerta, J., White, M.C., and Etzel, R.A. 1996 Effects of air pollution on the respiratory health of asthmatic children living in Mexico City. *Am. J. Respir. Crit. Care Med.*, **154**, 300–307.

Romieu, I., Meneses, F., Ruiz Velasco, S., Sienra-Monge, J.J., Huerta, J., White, M.C., Etzel, R.A. and Hernandez, M. 1997 Effects of intermittent ozone exposure on respiratory health of asthmatic children in Mexico City. *Archives on Environmental Health*, **52**(5), 368–376.

Romieu, I., Meneses, F., Sienra-Monge, J.J., Huerta, J., Ruiz-Velasco, S., White, M.C., Etzel, R.A. and Hernandez, M. 1995 Effects of urban air pollutants on emergency visits for childhood asthma in Mexico City. *American Journal of Epidemiology*, **141**, 546–553.

Romieu, I., Weitzenfeld, H. and Finkelman, J. 1990 Urban air pollution in Latin America and the Caribbean: Health perspectives. *World Health Statistics Quarterly*, **43**, 153–167.

Romieu, I., Weitzenfeld, H. and Finkelman, J. 1991 Urban air pollution in Latin America and the Caribbean: Health perspectives. *Journal of the Air Waste Management Association*, **41**, 1166–1170.

Saenger, P., Markowitz, M.E. and Rosen, J.F. 1984 Depressed excretion of 6B-hydroxycortisol in lead-toxic children. *Journal of Chinese Endocrinology and Metabolism*, **58**, 363–367.

Saldivar, P.H.N., Pope, C.A., Schwartz, J., Dockery, D.W., Lichtenfels, A.J., Salge, J.M., Barone, I. and Bohm, G.M. 1995 Air pollution and mortality in eldrely people: a time series study in São Paulo, Brazil. *Archives of Environmental Health*, **50**, 159–163.

Samet, J.M., Lambert, W.E., Skipper, B.J., Cushing, A.H., Hunt, W.C., Young, S.A., McLaren, L.C., Schwab, M. and Spengler, J.D. 1993 Nitrogen dioxide and respiratory illness in infants. *American Review of Respiratory Disease*, **148**, 1258–1265.

Samet, J.M., Speizer, F.E., Bishop, Y., Spengler, J.D. and Ferris, B.G. Jr. 1981 The relationship between air pollution and emergency room visits in an industrial community. *Journal of the Air Pollution Control Association*, **31**, 236–240.

Samet, J.M. and Utell, M.J. 1990 The risk of nitrogen dioxide: What have we learned from epidemiological and clinical studies. *Toxicology and Industrial Health*, **26**, 247–262.

Samet, J.M., Zeger, S.L. and Berhans, K. 1995 The association of mortality and particulate air pollution. *The Phase I Report of the Particle Epidemiology Evaluation Project*. Health Effects Institute, Cambridge, MA.

Sanin, L.H. 1995 Proteccion contra plomo durante el embarazo a traves del consumo de lecehe y jugo de naranja: El caso de la cuidad de Mexico. Paper presented at the International meeting "Plomo en las Americas". Instituto Nacional de Salud Publica. Cuernavaca, Mor, Mexico, 8–10 May.

SCAR 1984 *Report of the Scientific Review Panel on Benzene*. Staff of California Air Resource Los Angeles California Department of Health Services.

Schelegle, E.S., Siefkin, A.D. and McDonald, R.J. 1991 Time course of ozone-induced neutrophilia in normal humans. *American Review of Respiratory Disease*, **143**, 1353–1358.

Schwartz, J. 1989 Lung function and chronic exposure to air pollution: A cross-sectional analysis ofNHANES II. *Environmental Research*, **50**, 309–321.

Schwartz, J. 1991 Particulate air pollution and daily mortality. Paper presented at the Society for Occupational and Environmental Health. "Health Effects of Air Pollution: Impact of Clean Air Legislation". Crystal City, VA, March 25–27, 73.

Schwartz, J. 1994a Air pollution and daily mortality: A review and meta analysis. *Environmental Research*, **64**, 36–52.

Schwartz, J. 1994b What are people dying of on high air pollution days? *Environmental Research*, **64**, 26–35.

Schwartz, J. and Marcus, J. 1990 Mortality and air pollution in London; a time series analysis. *American Journal of Epidemiology*, **131**, 185–194.

Schwartz, J. and Otto, D.A. 1987 Blood lead, hearing threshold, and neurobehavioral development in children and youth. *Archives of Environmental Health*, **42**, 153–160.

Schwartz, J., Slater, D., Larson, T.V., Pierson, W.E. and Koenig, J.Q. 1993 Particulate air pollution and hospital emergency room visits for asthma in Seattle. *American Review of Respiratory Disease*, **147**, 826–31.

Seal, E. Jr., McDonnell, W.F., House, D.E., Salaam, S.A., DeWitt, P.J., Butler, S.O., Green, J. and Raggio, L. 1993 The pulmonary response of white and black men and women to six concentrations of ozone. *American Review of Respiratory Disease*, **147**, 804–810.

Sherwin, R.P. and Richters, V. 1990 Centriacinar region (CAR) disease in the lung of young adults. A preliminary report. Manuscript from Air and Waste Management Association Meeting, Los Angeles.

Shukla, R., Bornschein, R.L. and Dietrich, K.N. 1987 Effects of fetal and early postnatal lead exposure on child's growth in stature. The Cincinnati lead study. In: S.E. Lindberg and T.C Hutchinson [Eds] International conference: *Heavy Metals in the Environment*; New Orleans, LA, CEP Consultants, Ltd., Edinburgh, 210–212.

Shy, C.M. 1990 Lead in petrol: the mistake of the XXth century. *World Health Statistics Quarterly*, **43**, 168–176.

Silverman, D.T., Hoover, R.N., Albert, S. and Graff, K.M. 1983 Occupation and cancer of the low urinary tract in Detroit. *Journal of the National Cancer Institute*, **70**, 237–254.

Silverman, D.T., Hoover, R.N., Mason, K.J. and Swanson, G.M. 1986 Motor-exhaust-related occupation and bladder cancer. *Cancer Research*, **46**, 2113–2116.

Slade, R., Stead, A.G., Graham, J.A., Hatch, G.E. 1985 Comparison of lung antioxidant levels in humans and laboratory animals. *American Review of Respiratory Disease*, **131**, 742–6.

Smith, C.E., Koren, H.S., Graham, D.G. and Johnson, D.A. 1993 Mast cell tryptase is increased in nasal and bronchial alveolar lavage fluids of humans after ozone exposure. *Inhalation Toxicology*, **5**, 117–127.

Smith, M., Delves, T., Lansdown, R.N., Clayton, B. and Graham, P. 1983 The effects of lead exposure on urban children: The Institute of Child Health/ Southampton Study. *Developments in Medical Child Neurology*, Suppl., **47**, 1–54.

Speizer, F.E., Ferris, B. Jr., Bishop, Y.M. and Spengler, J.D. 1980 Respiratory disease rates and pulmonary function in children associated with NO_2 exposure. *American Review of Respiratory Disease* **121**, 3–10.

Spektor, D.M., Lippmann, M., Lioy, P.J., Thurston, G.D., Citak, K., James, D.J., Bock, N., Speizer, F.E. and Hayes, C. 1988 Effects of ambient ozone on respiratory function in active normal children. *American Review of Respiratory Disease*, **137**, 313–320.

Spengler, J.D., Brauer, M. and Koutrakis, P. 1990 Acid air and health. *Environmental Science and Technology*, **24**, 946–956.

Steenland, K. 1986 Lung cancer and diesel exhaust: A review. *American Journal of Industrial Medicine*, **10**, 177–189.

Stern, B. B., Jones, L., Raizenne, M., Burnett, R., Meranger, J.C. and Franklin, C.A. 1989 Respiratory health effects associated with ambient sulfates and ozone in two rural Canadian communities. *Environmental Research*, **49**, 20–39.

Stern, F.B., Halperin, W.E., Hornung, R.W., Ringenburg, V.L. and McCammon, C.S. 1988 Heart disease mortality among bridge and tunnel officers exposed to carbon monoxide. *American Journal of Epidemiology*, **128**, 1276–1288.

Stroup, N. E., Blair, A. and Erikson, G.E. 1986 Brain cancer and other causes of death in anatomists. *Journal National Cancer Institute*, **77**, 1217–1224.

Sunyer, J., Sáez, M., Murillo, C., Castellsague, J., Martínez, F. and Anto, J.M. 1993 Air pollution and emergency room admissions for chronic obstructive pulmonary disease: a 5-year study. *American Journal of Epidemiology*, **137**, 701–705.

Tepper, J.S., Costa, D.L., Lehmann, J.R., Weber, M.F. and Hatch, G.E. 1989 Unattenuated structural and biochemical alterations in the rat lung during functional adaptation to ozone. *American Review of Respiratory Disease*, **140**, 493–501.

Thurston, G.D., Ito, K. and Hayes, C.G. 1989 Re-examination of London, England. Mortality in relation to exposure to acid aerosols during 1963–1972 winters. *Environmental Health Perspectives*, **79**, 73–82.

Thurston, G.D., Ito, K., Kinney, P.L. and Lippmann, M. 1992 A multi-year study of air pollution and respiratory hospital admissions in three New York state metropolitan areas: results for 1988 and 1989 summers. *Journal of Exposure Analysis and Environnental Epidemiology*, **2**, 429–450.

Thurston, G.D., Ito, K., Hayes, C.G., Bates, D.V. and Lippman, M. 1994 Respiratory hospital admissions and summertime haze air pollution in Toronto, Ontario: consideration of the role of acid aerosols. *Environmental Research*, **65**, 271–290.

Touloumi, G and Katsouyanni, K. 1995 Short term effect of air pollution on mortality: Results of the APHEA project for the Athens population. *Epidemiology*, **6**, S59.

Tuppurainen, M., Wagar, G., Kurppa, K., Sakari, W., Wambugu, A., Froseth, B., Alho, J. and Nykyri, E. 1988 Thyroid function as assessed by routine laboratory tests of workers with long-term exposures. *Scandinavian Journal of Work, Environment and Health*, **14**, 175–180.

US EPA 1982a *Air Quality Criteria for Oxides of Nitrogen.* Report 600/8–82–026F, United States Environmental Protection Agency, Washington D.C.

US EPA 1982b *Air Quality Criteria for Particulate Matter and Sulfur Oxides.* Report 600/8–82–029C, United States Environmental Protection Agency, Washington D.C.

US EPA 1983 *Revised Evaluation of Health Effects Associated with Carbon Monoxide Exposure: An Addendum to the 1979 Air Quality Criteria Document for Carbon Monoxide.* Report 600/8–83–033a, United States Environmental Protection Agency, Washington D.C.

US EPA 1985 *Interim Quantitative Cancer Unit Risk Estimates due to Inhalation of Benzene.* Internal Report No. 600/X–85–022, Carcinogenic Assessment Group, United States Environmental Protection Agency, Washington D.C.

US EPA 1986a *Air Quality Criteria for Ozone and Other Photochemical Oxidants*, Volume V, Report 600/8–84/020eF, United States Environmental Protection Agency, Washington D.C.

US EPA 1986b *Air Quality Criteria for Lead, June 1986 and Addendum September 1986*, Report 600/8–83–018F, Research Triangle Park, N.C. and Office of Research and Development, Office of Health and Environmental Assessment, Environmental Criteria Assessment Office, United States Environmental Protection Agency, Washington D.C.

US EPA 1987 *Revision to the National Ambient Air Quality Standards for Particulate Matter: Final Rules.* Fed. Regist. 52 (126), 24634–69, United States Environmental Protection Agency, Washington D.C.

Utell, M. 1994 Public health risk from motor vehicle emissions. *Annual Review of Public Health*, **15**, 157–178.

Van De Lende, R., Kok, T.J., Reig, R.P., Quanjer, P.H., Schorten, J.P. and Orie, G. 1981 Decreases in VC and FEV_1 with time: indicators for effects of smoking and air pollution. *Bulletin European De Physiopathologie Respiratoire*, **17**, 775–792.

Van Raalte, H.G.S. and Grasso, P. 1982 Hematological, myelotoxic, clastogenic, carcinogenic, and leukemogenic effects of benzene. *Regulatory Toxicology Pharmacology*, **2**, 153–176.

Vander, A.J. 1988 Chronic effects of lead on the renin-angiotensin system. *Environmental Health Perspectives*, **78**, 85–89.

Vaughan, T.L., Strader, C., Davis, S. and Daling, J.R. 1986a Formaldehyde and cancers of the pharynx, sinus and nasal cavity, I. Occupational exposures. *International Journal of Cancer*, **38**, 677–683.

Vaughan, T.L., Strader, C., Davis, S. and Daling, J.R. 1986b Formaldehyde and cancers of the pharynx, sinus, and nasal cavity, II. Residential exposures. *International Journal of Cancer*, **38**, 685–688.

Vedal, S., Schenker, M.B., Muñoz, A., Samet, J., Batterman, S. and Speizer, F.E. 1987 Daily air pollution on children´s respiratory symptoms and peak expiratory flow. *American Journal of Public Health*, **77**, 694–698.

Walrath, J. and Fraumeni, J.F. 1983 Proportionate mortality among New York embalmers. In: J.E Gibson. [Ed.] *Formaldehyde Toxicity*. Hemisphere, New York, 227–236.

Walrath, J. and Fraumeni, J.F. 1984 Cancer and other causes of death among embalmers. *Cancer Research*, **44**, 4638–4641.

Ward, N.I., Watson, R. and Bryce-Smith, D. 1987 Placenta element levels in relation to fetal development for obstetrically normal births: A study of 37 elements, Evidence for the effects of cadmium, lead, and zinc on fetal growth and for smoking as a source of cadmium. *International Journal of Biosocial Research*, **9**, 63–81.

Ware, J.H., Ferris, B.G. Jr., Dockery, D.W., Spengler, J.D., Stram, D.O. and Speizer, F.E., 1986 Effects of ambient sulfur oxides and suspended particles on respiratory health of preadolescent children., *American Review of Respiratory Disease*, **133**, 834–842.

White, M.C., Etzel, T.A., Wilcox, W.D. and Lloyd, C. 1994 Exacerbations of childhood asthma and ozone pollution in Atlanta. *Environmental Research*, **65**, 56–68.

White, M.C. and Etzel, R.A. 1991 Childhood asthma and ozone pollution in Atlanta. Paper presented at the Society for Occupational and Environmental Health meeting on Health Effects of Air pollution: Impact of Clean Air Legislation, 25–27 March 1991, Crystal City, VA.

Whittemore, A. and Korn, E.L. 1980 Asthma and air pollution in the Los Angeles area. *American Journal of Public Health*, **70**, 687–696.

WHO 1977 *Oxides of Nitrogen*. Environmental Health Criteria No. 4, World Health Organization, Geneva.

WHO 1978 *Photochemical Oxidants*. Environmental Health Criteria No. 7, World Health Organization, Geneva.

WHO 1979 *Carbon Monoxide*. Environmental Health Criteria No. 13, World Health Organization, Geneva.

WHO 1987a Nitrogen Dioxide. In: *Air Quality Guidelines for Europe*. WHO Regional Publications, European Series No. 23, World Health Organization Regional Office for Europe, Copenhagen, 297–314.

WHO 1987b Ozone and other photochemical oxidants. In: *Air Quality Guidelines for Europe*. WHO Regional Publications European Series No. 23, World Health Organization, Regional Office for Europe, Copenhagen, 315–326.

WHO 1987c Sulfur dioxide and particulate matter. In: *Air Quality Guidelines for Europe*. WHO Regional Publications, European Series No. 23, World Health Organization, Regional Office for Europe, Copenhagen, 338–360.

WHO 1987d Carbon Monoxide. In: *Air Quality Guidelines for Europe*. WHO Regional Publications, European Series No. 23, World Health Organization, Regional Office for Europe, Copenhagen, 210–220.

WHO 1987e Lead. In: *Air Quality Guidelines for Europe*. WHO Regional Publications, European Series No. 23, World Health Organization, Regional Office for Europe, Copenhagen, 242–261.

WHO 1987f Benzene. In: *Air Quality Guidelines for Europe*. WHO Regional Publications, European Series No. 23, World Health Organization, Regional Office for Europe, Copenhagen, 45–58.

WHO 1987g Polycyclic aromatic hydrocarbons (PAH's). In: *Air Quality Guidelines for Europe*. WHO Regional Publications, European Series No. 23, World Health Organization, Regional Office for Europe, Copenhagen, 105–117.

WHO 1989 *Formaldehyde*. Environmental Health Criteria No. 89, World Health Organization, Geneva.

WHO 1993 *Benzene*. Environmental Health Criteria No. 150, World Health Organization Geneva.

WHO 1995a *Updating and revision of the air quality guidelines for Europe, Meeting of the WHO Working Group "Classical" Air Pollutants.* EUR/ICP/EHAZ 94 05/PB01, World Health Organization, Regional Office for Europe, Copenhagen.

WHO 1995b *Updating and revision of the air quality guidelines for Europe, Report on the WHO Working Group on Inorganic Air Pollutants.* EUR/ICP/EHAZ 94 05/MT04, World Health Organization, Regional Office for Europe, Copenhagen.

WHO 1996a *Diesel Fuel and Exhaust Emissions*. Environmental Health Criteria No. 171, World Health Organization, Geneva.

WHO 1996b *Updating and revision of the air quality guidelines for Europe, Report on a WHO Working Group on Volatile Organic Compounds.* EUR/ICP/EHAZ 94 05/MT12, World Health Organization, Regional Office for Europe, Copenhagen.

Winneke, G., Krämer, U., Brockhaus, A., Ewers, U., Kujanek, G., Lechner, H. and Janke, W. 1983 Neuropsychological studies in children with elevated tooth-lead concentration, Part II Extended study. *International Archives of Occupational and Environmental Health*, **51**, 231–252.

Winneke, G., Beginn, U., Ewert, T., Havestadt, C., Krämer, U., Krause, C., Thron, H.L. and Waner, H.M. 1985 Comparing the effects of perinatal and later childhood lead exposure on neuropsychological outcome. *Environmental Research*, **38**, 155–167.

Wong, O., Morgan, R.W., Kheifets, L., Larson, S.R. and Whorton, M.D. 1985 Mortality among members of a heavy construction equipment operators union with potential exposure to diesel exhaust emissions. *British Journal of Industrial Medicine*, **42**, 435–448.

Yin, S.N., Li, G.L., Tain, F.D., Fu, Z.I., Chen, Y.J., Luo, S.J., Ye, P.Z., Zhang, J.Z. and Wang, G.C. 1989 A retrospective cohort study of leukemia and other cancers in benzene workers. *Environmental Health Perspectives*, **82**, 207–213.

Ziegler, E.E., Edward, B.B., Jensen,R.L., Mahaffey, K.R. and Fomon, S.J. 1978 Absorption and retention of lead by infants. *Pediatric Research*, **12**, 29–34.

Zmirou, D., Balducci, F., Barumandzadeh, T., Ritter, P., Laham, G. and Ghilardji, P. 1995 Daily mortality and air pollution in Lyon, 1985–1990, A European Approach. *Epidemiology*, **6**, S61.

Chapter 3*

HEALTH EFFECTS FROM ROAD TRAFFIC NOISE

Sounds provide an essential contact between humans and the surrounding world. Awareness of familiar sounds such as the waves at sea, footsteps of family members and music, induces reactions of recognition, pleasure and satisfaction. The harmony between normal sounds and periods of quietness is also important, particularly for music. Other sounds such as creaking floors in an empty house, sudden bangs and unpleasant music, induce alertness, fear and annoyance. Such sounds are generally referred to as noise, although this is a subjective assessment. Definitions of, as well as adverse reactions to, different sounds vary between individuals as a result of experience, attitudes and knowledge.

Sound has particular characteristics which distinguish it from other kinds of environmental pollutants, such as chemical agents. Sound, or noise, is part of every day life and is necessary for the normal functioning of the human body. Persons who are kept under conditions of complete silence may develop symptoms of mental disorder. The absence of auditory communication due to deafness may induce personality changes, and conditions of complete silence can be experienced as frightening. As sound levels increase in intensity above a certain level, negative effects become more dominant and the ultimate effect is direct physical trauma when the receptor organ is destroyed. Although positive reactions associated with sound exposure are important, it is usually the negative effects that attract attention.

3.1 Environmental noise

Noise has always been an important environmental problem for people. In ancient Rome, rules existed governing the noise emitted from the ironed wheels of wagons which battered the stones on the pavement, causing disruption of sleep and annoyance to the inhabitants of the city. In Medieval Europe, horse carriages and horseback riding were not allowed at night in certain cities to ensure a peaceful night's sleep for the city dwellers.

* *This chapter was prepared by Ragnar Rylander*

However, the noise problems of the past are incomparable with those of modern society. An immense number of motor cars constantly travel through our cities and the countryside and heavily-laden lorries with diesel engines, which have been badly silenced for engine and exhaust noise, are present in cities and on highways, day and night. Aircraft and trains add to the environmental noise problem. In industry, machinery emits high noise levels and amusement centres and pleasure vehicles impinge upon leisure time relaxation.

In comparison with other pollutants, the control of environmental noise has been hampered by insufficient knowledge of its effects on humans and of dose-response relationships, as well as by a lack of defined criteria. Although it has been suggested that noise pollution is primarily a "luxury" problem for developed countries, exposure is often higher in developing countries, due to bad planning and to the poor construction of buildings. The effects of noise in developing countries are just as widespread as in developed countries and the long-term consequences for health are the same. Practical actions to limit and control the exposure to environmental noise are therefore essential. Such action must be based upon proper scientific evaluation of available data on effects, and particularly on dose-response relationships. The basis for this is the process of risk assessment as described below.

A strategy for evaluating the impact of an exposure to noise, can suitably borrow terminology from the field of toxicology. The broad integrated process of risk assessment, can be described by the following elements:

- *Dose description.* The level of the agent must be described in a way that is appropriate with regard to the effects observed.
- *Hazard identification.* This is the qualitative description of how an agent may adversely effect human health or well being. It may represent a very wide assessment of possible risks, using a variety of experimental and epidemiological studies.
- *Hazard assessment.* The qualitative and quantitative evaluation of the nature of adverse effects and their expression as functions of exposure (dose-response relationship).
- *Risk estimation.* The integration of hazard assessment and dose description to quantify the risk to be accepted by the community (guidelines, threshold values).

A "hazard" signifies the potential of a specific agent to cause harm, and is an inherent property of the agent *per se*, in terms of its toxicity. A "risk" represents the quantitative statement of the probability of occurrence of a defined adverse effect.

3.2 Dose description

Sound is a longitudinal wave motion which occurs when a sound source produces compression and rarefaction of air volumes. The wave gradually spreads to air volumes further away from the source. Sound travels through the air at a velocity of approximately 340 m s^{-1}. The physical quantity is the sound pressure amplitude p measured in Pascal (Pa). The audible sound covers a large range of pressures from 0.00002 Pa at the threshold of hearing to 20 Pa at the threshold of pain. It would be impractical to work with this large range of numbers and so an artificial quantity, the sound pressure level, L_p , has been created:

$$L_p = 20\log_{10}\left(\frac{p}{p_{ref}}\right)[\mathrm{dB}]$$

The reference pressure p_{ref} corresponds to the audition threshold and has an internationally agreed value of 2×10^{-5} N m^{-1} but is often given in micro Pascal (i.e. 20 μPa). The sound pressure level is given in decibel units (dB). Noise from different sources combines to produce a sound pressure level higher than that from any individual source. Two equally intense sound sources operating together, produce a sound-pressure level which is 3 dB higher than one alone and 10 sources produce a 10 dB higher sound level. The dB values cannot be directly added because they are logarithmic quantities. Apart from the physical intensity, sounds contain many tones at different frequencies. The soundwave's frequency expresses the number of vibrations per second in units of Hertz (Hz). Sound exists over a very wide frequency range. Audible sound for young people lies between 16 Hz and 20,000 Hz. Sound with frequencies below 16 Hz (normally inaudible) is called infrasound and sound over 20,000 Hz, which is also normally inaudible, is called ultrasound. Low frequency sounds are not strictly defined, but are generally referred to as frequencies from 16 to 250 Hz.

When measuring sound pressure level, an instrument which mimics the varying sensitivity of the ear to the sound of different frequencies is usually used. This is achieved by building a filter, with a similar frequency response to that of the ear, into the sound intensity recording instrument. This filter is known as an A-weighted filter. Measurements of sound pressure level made with this filter are called A-weighted sound level measurements, and the unit is dB(A). The dB(A) levels for some common sounds in our environment are given in Table 3.1.

In addition to the above described frequency-weighted filter, more complicated measures have been developed to describe the human response to

Table 3.1 Approximate noise levels arising from common sounds

Sources	Noise level (dB(A))
Close to jet engine	130
Rock drilling machine	120
Pop concert	110
Heavy lorry	90
Passenger car	75
Normal conversation	65
Quiet suburban street	55
Threshold for sleep disturbance	45
Very quiet room	30
"Uncomfortably" quiet	15
Hearing threshold	0

complex sounds. One such unit is the Perceived Noise Level (PNL) which is based on frequency-weighted sound levels measured in successive 0.5 seconds intervals during the occurrence of a sound. Different sound spectra for noise rating have also been developed. The sound spectrum curves, i.e. noise rating (NR) curves, serve as a frame of reference for rating noise environments. Such curves were originally evolved for the rating of outdoor community noise and are present as a recommendation of the International Organization for Standardization (ISO, 1982).

From an acoustic point of view, environmental noise is often a complex phenomenon. This is illustrated for road traffic noise in Figure 3.1. At a specific site along a road, the noise level varies with time depending upon the type of vehicle that passes. With a small number of vehicles, the noise level will return to the background level between passages, whereas a larger number of vehicles will turn the situation into one of almost continuous noise, fluctuating between the levels generated by particularly noisy vehicles such as lorries and the lower levels generated by cars. This complex acoustic pattern is expressed traditionally as the summation of sound pressure level over a certain period of time. Various methods of calculating an average have been developed. The exposure to noise from various sources is most commonly expressed as the A-weighted average sound pressure level over a specific time period T, such as 24 hours. This gives a value of the equivalent continuous sound pressure level L^{T}_{eq} in units of dB(A) which is derived from the following mathematical expression:

$$L_{eq}^{T} = 10\log_{10}\left[\left(\frac{1}{T}\right)\int_{0}^{T} 10^{\frac{L_P(t)}{10}}\, \mathrm{dt}\right] \mathrm{dB(A)}$$

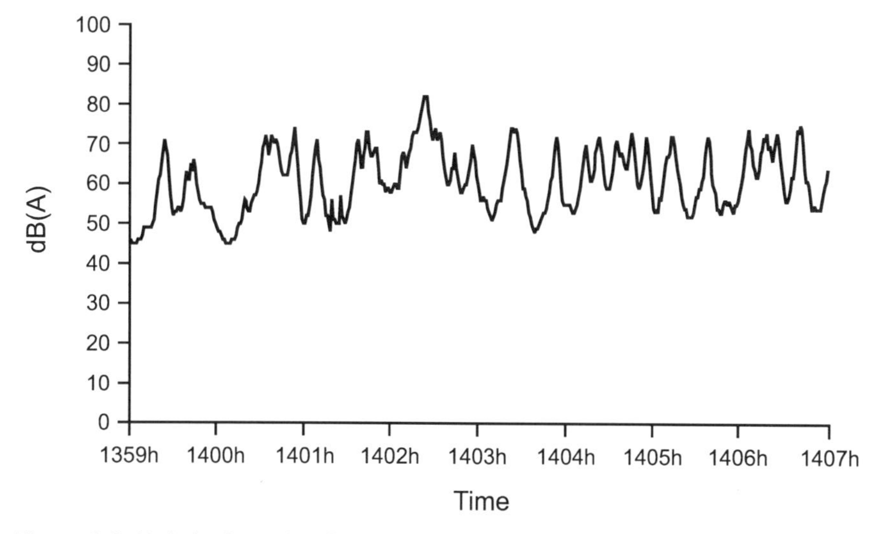

Figure 3.1 Variation in road traffic noise with time. Data provided by ECOTOX, Geneva

The integral is a measure of the total sound energy during the period *T*. Thus, L^{T}_{eq} is the level of that steady sound which, over the same interval of time *T* as the fluctuating sound of interest, has the same effective sound pressure. For practical calculations of the equivalent sound pressure level the integral in the equation is replaced by a sum. This calculation is a physical concept that implies that the same average level of chosen time can consist either of a larger number of events with relatively low levels, or of fewer events with high levels. It does not necessarily agree with common experience on how environmental noise is experienced, nor with the neuro-physiological characteristics of the human being.

Another method of calculation is the use of the statistical distribution level such as the noise levels L_{01}, L_{10} or L_{50}. These are also based on an average value of events and noise levels, and the index refers to the proportion of time for which the specific noise level is reached, e.g. L_{10} is the level in dB(A) during 10 per cent of the measuring period. A complete review of all noise indices can be found in relevant handbooks or criteria documents (e.g. Berglund and Lindvall, 1995).

The concept of average level has two critical features. A few events with a high noise level will have the same L_{eq} as a large number of events at a lower noise level. From a biological point of view, it is unlikely that these two noise scenarios will cause an equal effect in the exposed populations. A second critical feature for the average noise value relates to the number of events. If the L_{eq} value for a certain number of cars at a certain distance from the road is

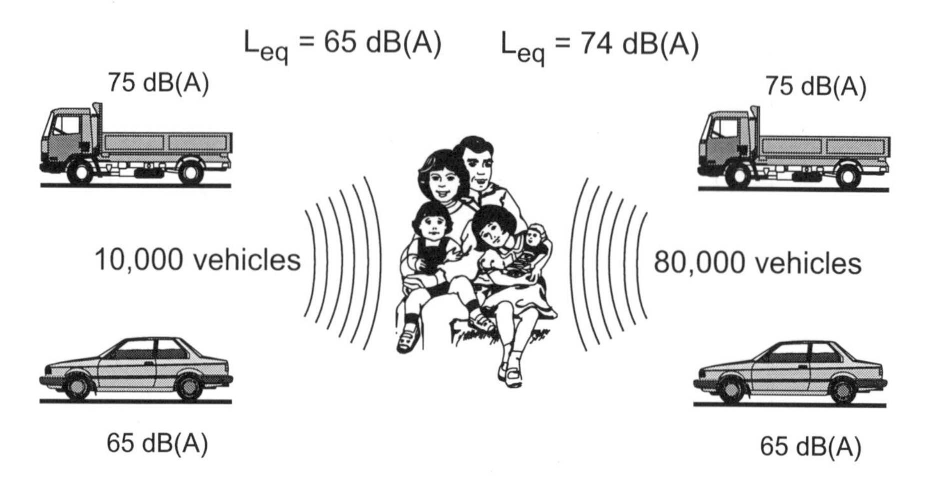

Figure 3.2 Maximum noise levels (L_{eq}) from cars and lorries form two different total numbers of cars and lorries but with the same proportion of lorries. Redrawn with permission from WHO

for example 65 dB(A), then at that distance passing cars will each cause a maximum noise level of 65 dB(A) and a few noisy lorries will reach 75 dB(A). If the number of cars increases, the L_{eq} value will gradually increase, although the noise level from each passing car is still 65 dB(A) and 75 dB(A) from a single lorry remains the noisiest event. Under extreme circumstances this may even imply that the L_{eq} value of 65 dB(A) is present at a site, far away from the road where the noise levels from individual vehicles (even very noisy ones) is very low (Figure 3.2). The critical factor in the situation described above is the number of events. This poses the question: to what extent is the number of events related to the human perception of environmental stimuli?

Biological receptor systems (such as hearing, seeing and feeling) are designed to discriminate between variations in exposure. Noise arouses the receptor organs, not through the total amount of energy but through density per unit of time (intensity) combined with frequency characteristics. With a large number of events, however, the individual event can no longer be discriminated (e.g. flickering of light or applying pressure on the skin with high frequency). It is thus plausible that an increase in the number of noise events above a certain number, will not lead to an increased effect in the exposed population. Furthermore, the major part of human reactions following exposure to environmental noise stems from the interpretation of the noise at the cerebral level. An example is the intense fright reaction caused by noise of an unknown character occurring at night but where the noise level is very low. By contrast, the comparatively high noise level of a passing train can be experienced without any adverse reaction at all. From a biological point of view, it is thus unlikely that the appropriate human noise dose consists of an average exposure over time. A biologically relevant dose should be based on neurophysiological receptor principles and thus probably comprises some of the noise events (excluding others without biological significance) as well as the level of these events.

3.3 Hazard assessment and risk estimation

Noise may cause physical, physiological and psychological effects in humans. Sound waves act physically against the ear drum with subsequent risk for hearing impairment, threshold shift and hearing loss, or they interfere with ototoxic drugs and environmental chemicals. Physical effects may be referred to as direct effects. Through the nerve impulse to the central nervous system, noise may induce physiological changes and may finally register cognitively and cause psychological damage. The effects that are measured in the exposed human may be a discrete physiological reaction or a complex reaction, such as sleep disturbance or an effect on performance. Physiological and psychological effects are referred to as indirect effects. For all effects, those appearing after a single or rare exposure as well as those occurring after repeated exposures (chronic exposure), need to be considered.

3.3.1 Physical effects

Noise of a high enough level will cause temporary or permanent damage to the hearing organ. The mechanisms behind these injuries are well understood and dose-response relationships have been presented for continuous as well as intermittent exposure. High level noise giving rise to noise-induced hearing deficits can occur in occupational situations, open air concerts, discotheques, motor sports events, shooting ranges, and from loudspeakers or headphones within dwellings. Due to the considerable variation in human sensitivity to noisy environments and to their likelihood of causing hearing impairment, the hazardous nature of a noisy environment is described in terms of "damage risk". This is defined as the probability, in a noise exposed population, of suffering from noise-induced hearing loss. This risk is considered to be negligible at equivalent noise exposure levels below 75 dB(A) over an exposure period of 8 hours (Berglund and Lindvall, 1995). With respect to environmental noise and road traffic noise, no risk for hearing damage exists. Levels in the general environment do not reach those which will induce hearing damage, even close to traffic along heavily congested streets.

The interaction between noise from road traffic and other sounds in the environment can have important effects. Levels causing speech interference are often present outside, and also within, buildings close to heavily congested roads. Vulnerable groups in the population are school children in noisy classrooms. One study evaluated the influence of traffic noise (related to speech interference) on the reading ability of children in New York (Cohen *et al.*, 1973). It was found that children living on the lower levels of buildings and hence exposed to more road traffic noise, showed greater impairment of auditory discrimination and reading achievement than children living on the higher levels.

People with hearing deficiencies are also at risk from exposure to noise. Aniansson *et al.* (1983) studied speech interference, annoyance and changes in mood in groups of people with different degrees of hearing deficiency. These groups were exposed to 45 dB(A) and 55 dB(A) traffic noise in a laboratory where they had to perform four everyday activities. The major finding was a higher rating of annoyance among men with noise-induced hearing loss compared with men with normal hearing. This effect was related to activities in which noise interfered with speech. To achieve good speech intelligibility for persons with impaired hearing due to age and/or noise, it was recommended that the noise level outdoors should not exceed 50 dB(A) (Aniansson and Peterson, 1983). This value took into account the reduction in noise, by about 25 dB(A) between outdoor and indoor situations, that is generally observed in Scandinavian countries.

According to a recent publication on community noise prepared from a meeting of a WHO task group (Berglund and Lindvall, 1995), communication interference starts at sound levels of about 50 dB(A). A speech given with background noise levels of about 45 dB(A) is 100 per cent intelligible, but can still be understood fairly well with background noise levels of 55 dB(A). Speech communication is affected by reverberation time. Even in quiet environments a reverberation time of below 0.6 seconds is desirable.

3.3.2 Physiological effects

The immediate response to a noise stimulus comprises a startle and a defence reaction. The startle response is a reflex resulting in contraction of muscles around the eyes, in the limbs and the eyelids (Thachray, 1972). The response causes an attachment of consciousness to the noise and its source, often followed by an orientation towards the noise source through involuntary muscular movements. This is a typical reaction which occurs after exposure to unexpected or sudden noises, irrespective of their physical noise level. After an interpretation of the noise in higher centres of the brain, a defence and fright reaction may follow. The startle reflex is accompanied by an increase in blood pressure and pulse frequency of a very short duration (up to 30 seconds) and, in extreme situations, an increased secretion of stress hormones.

Habitation to the startle reflex is very poor and the response can be induced repeatedly, over the same day and over longer time periods (Rylander and Dancer, 1978). With environmental noise, such reactions occur when sudden noise levels exceed the normal level, such as when a motor cycle without silencers drives down a road, a heavy lorry passes in the middle of the night or an unusually noisy aircraft flies over the house.

Based upon observations of an increase in blood pressure after acute exposure to noise, it has been suggested that long-term exposure to noise could cause a persistent increase in blood pressure. This hypothesis can be evaluated by examining experiences from epidemiological studies. The results of studies made on persons exposed to high levels of noise in industry are conflicting. A review of the cardiovascular effects of noise reported that 55 studies had been performed on the relationship between noise and blood pressure and about 80 per cent of these had reported some form of positive association (Dejoy, 1984) who noted that *"a paucity of quantitative data ... makes it difficult to assess the strength of association or to derive a dose-response relation"*. In a study on the effects of industrial noise on the prevalence of hypertension, Yiming *et al.* (1991) studied a group of 1,101 female workers in a textile mill in Beijing, China. Essentially, the group had worked in different workshops in this mill for all their working lives and all had worked for at least five years. The noise levels within the plant were assessed and appear to have been constant since 1954, resulting in well defined noise exposures for these workers. A cross-sectional study design was used in which blood pressures were determined and questionnaires administered to the workers over a two month period. As well as demographic information, data were gathered on personal and family history of hypertension, current use of prescription drugs, alcohol, tobacco, and salt in the diet. Logistic regression indicated that exposure to noise was a significant determinant of prevalence of hypertension, but third in order of importance behind family history of hypertension and use of salt. Cumulative exposure to noise was not an important dose-related variable, suggesting that for those susceptible to the effect, hypertension was manifested within the first five years of exposure.

A few studies have also been made in the general population, comparing those living along noisy streets with those living near quiet streets. The results from some early studies show slightly higher blood pressure among people living along roads with heavy traffic compared with those living in quiet areas (Knipschild and Sallé, 1979). In those studies other risk factors for cardiovascular disease, such as food intake or smoking, were not controlled for although this has been done in more recent studies. In a study designed to study risk factors for heart disease, a total of 5,000 men was selected at random in two cities (Babish *et al.*, 1993). Noise maps were made in the areas and measurements performed of road traffic noise exposure. Clinical examinations included blood pressure, blood biochemistry, plasma lipids and lung function. Questionnaires were used to obtain data on general health, smoking and social class. The examinations were performed on two occasions in 1979–83 and 1984–88. Using logistic regression techniques, several risk factors for cardiovascular disease were controlled in the analysis. No significant risk increases could be related to the higher traffic noise exposure.

The results from these studies suggest that environmental noise exposure is not an important risk factor for cardiovascular disease. If a persistent increase in blood pressure can be induced by environmental noise, it is likely that it is modulated through the basic reflex functions described earlier. This effect could not be large, and it would be difficult to distinguish the influence of noise from other environmental stress factors which also produce a slight increase in blood pressure.

3.3.3 Sleep disturbance

Exposure to noise can induce disturbance of sleep by causing difficulty in falling asleep, alterations in sleep rhythm or depth, and through being woken up. An objective recording of sleep can be obtained by measuring the electrical activity of the brain as an electroencephalogram (EEG). This requires that subjects carry several electrodes on their heads and that they are connected to a recording device, either directly or via telemetry.

An abundance of information exists on EEG-recorded sleep in laboratory and field conditions. There is a fairly uniform consensus that changes in the depth and pattern of sleep represent acute effects following exposure to noise at levels of 45–50 dB(A) peak levels. A good example of a sleep study is that of an investigation from France in which sleep disturbance was studied among a population living near a railway and a major road (Vernet, 1979). Two areas with similar noise exposures were selected and 10 persons of both sexes were studied in each area. Sleep patterns were registered using EEG and EMG (electromyogram) recordings transmitted to a mobile van outside the subjects' homes. A good correlation was found between EEG-defined sleep disturbance and peak noise level, but no correlation was found between the duration of noise and sleep disturbance. The percentage of responses increased when peak noise levels increased above 45 dB(A). Subjects were not woken up by noise below 52 dB(A) peak level. About 25 per cent of noise events at 70 dB(A) caused sleep interference. There were three times as many disturbances by road traffic noise as by railway noise for the same L_{eq} value.

Recording of body movements has also been used to study the effects of sleep and good correlations have been found between EEG changes and actimetry readings (wrist movements transmitted from a sensor, about the same size as a large wristwatch, to a recording device). Available information demonstrates that noise exposure induces changes in sleep pattern or causes additional body movements. Considerably less information exists on the medical consequences of disturbed sleep. Long-term effects of noise-induced sleep interruption could be subjective fatigue, changes in performance and subjective mood changes.

In a laboratory study, the effects of lorry noise with a maximum 60 dB(A) were studied during a period of two weeks (Öhrström *et al.*, 1988). The study was designed to illustrate habituation effects on body movements and heart rate as well as delayed effects, such as subjective sleep quality, state of mind and performance. Two different test groups, consisting of 24 persons who were sensitive or non-sensitive to noise, were studied. Their sleep was also registered in their homes during one week before and one week after their stay in the laboratory. After two nights, sleep quality was just as good in the laboratory as at home. The study showed that following the noise events there was an increase in heart rate by an average of 1.8 beats per minute for sensitive and 1.1 beats per minute for non-sensitive persons, and that the number of movements increased by three times after noise for sensitive persons compared with an increase of 2.5 times for non-sensitive persons. These reactions showed no sign of habituation. Subjective sleep quality was impaired for the sensitive group and had not improved at the end of the two-week period. This impairment (significant for non-sensitive persons) was still there at the end of the test period. Moreover, an increase in fatigue (mostly in the non-sensitive persons) and a reduced degree of extroversion (mostly for the sensitive persons) could be registered towards the end of the period of noise exposure. Both groups exhibited reduced activity and both groups performed worse in performance tests; this deterioration tended to continue towards the end of the study period.

On the basis of experience gained from these studies, a field study was carried out close to a road with heavy traffic (L_{eq} approximately 71 dB(A)) and in a quiet area far from the road (L_{eq} approximately 50 dB(A)) in order to observe the long-term effects of noise exposure during sleep (Öhrström, 1989). The study comprised a total of 106 persons who were interviewed about sleep quality, fatigue, mood and various medical and psycho-social symptoms. They were also asked separate questions about sleep and mood on three successive days. In the area close to the road with heavy traffic, the study showed a deterioration in sleep quality and mood, and a higher frequency of other symptoms such as tiredness, headache and nervous stomach. Those who described themselves as sensitive to noise reported poorer sleep quality and more symptoms. Among persons living along noisy roads, instant improvements in subjective sleep quality have been reported following the noise-insulation of windows (Wilkinson *et al.*, 1980; Öhrström and Björkman, 1983).

Studies on sleep interference due to noise agree that the most important exposure parameter is peak noise during night hours. Effects are generally reported at peak levels from 45 dB(A) indoors and therefore large parts of populations in cities are regularly exposed to noise levels which interfere with their sleep.

In conclusion, available evidence suggests that sleep disturbance is one of the major effects of environmental noise and that it may have serious adverse effects on normal functioning and health in exposed persons. According to the findings of a WHO task group measurable sleep disturbance effects start at equivalent continuous sound pressure levels of about 30 dB(A) (Berglund and Lindvall, 1995) and increase with increased maximum noise level. Even if the total equivalent noise level is fairly low, a small number of noise events with a high maximum sound pressure level will affect sleep. It is especially important to consider the noise events exceeding 45 dB(A) when the background level is low.

3.3.4 Psychological effects

Noise can be bothersome and can give rise to psychological and psychosomatic symptoms in the form of headaches, fatigue and irritability. Biochemical reactions indicating a general stress effect from noise have also been reported in animal and human studies (Cantrell, 1974).

In view of the information available on sleep disturbance and on the stress reaction occurring after noise exposure, psychiatric symptoms or disorders have received particular attention. Psychiatric effects could occur in three different ways: symptoms could develop among previously normal persons, their development could be accelerated in predisposed persons, or symptoms could appear temporarily under particular conditions.

No data are available on psychiatric disease and road traffic noise, but some studies on aircraft noise have been published. Epidemiological evidence for a relationship between aircraft noise exposure and psychiatric illness was presented by Abey-Wickrama *et al.* (1969). They described an increased admission rate to hospitals among a population living in an area exposed to high levels of noise. This observation prompted additional studies which employed more precise epidemiological techniques. A subsequent study by Gattoni and Tarnopolsky (1973) did not confirm Abey-Wickrama's findings. A further study by Jenkins *et al.* (1979) on psychiatric hospital admission rates in the same area over a four-year period also failed to demonstrate a higher admission rate among noise exposed persons. The latter authors made a careful analysis of possible reasons for the contradictory results and convincingly demonstrated the various shortcomings in the first study which could have accounted for the increase in admission rate observed in noise-exposed areas. Consequently, no epidemiological evidence is at present available which indicates that there is an increased risk for psychiatric disorders among general populations exposed to noise.

In another study on the effect of aircraft noise on mental health, Tarnopolsky *et al.* (1980) investigated 6,000 persons living in areas of London with different levels of aircraft noise. The subjects answered questionnaires relating to general health, psychiatric illness and annoyance. Of the noise-related effects, depression, irritability, awakenings and difficulty in falling asleep were significantly more frequent in the high noise area. The prevalence of symptoms was significantly higher among persons who expressed that they were annoyed by the noise. There was no relationship between the number of possible psychiatric cases and noise exposure. No increase in consumption of psychotropic drugs or use of medical services was found in the high noise area.

It could be hypothesised, according to previous studies on noise effects (e.g. Öhrström, 1989), that physical and psycho-social symptoms and reduced work capacity may occur as an effect of general annoyance and sleep disturbances caused by noise exposure. These symptoms may, however, also be dependent upon other circumstances such as chronic illness or difficulties in the family situation or work conditions. The individual capacity to handle stress might also be of importance in the development of different symptoms.

Öhrström (1991) performed a field survey in a quiet area and in an area exposed to an L_{eq} level of 72 dB(A) in order to elucidate possible psycho-social effects of road traffic noise. A questionnaire was constructed to evaluate not only annoyance reactions and sleep disturbance effects by noise, but also more long-term effects on psycho-social well-being (PSW). Psycho-social well-being was evaluated by 26 questions concerning depression, relaxation, activity, passivity, general well-being and social orientation. The postal questionnaire was answered by 151 persons in the quiet area and 118 persons in the noisy area, of which 97 lived in apartments facing the street and 21 persons in apartments facing the courtyard. The results showed that a higher proportion of those who lived in the noisy area in apartments with windows facing the street felt depressed. However, those persons in the noisy area who had windows facing the courtyard, were not more depressed than those who lived in the quiet area. Psycho-social well-being, especially depression and elation, was related to annoyance about noise.

In conclusion, the results available so far do not indicate that environmental noise provokes psychiatric disease. Noise may, however, act as a stressor, inducing symptoms among sensitive individuals. The exact conditions under which these individuals become vulnerable are not known, but it is conceivable that other environmental strains could act synergistically with noise.

3.4 Annoyance

Exposure to environmental noise may interfere with ongoing activities and can be experienced as bothersome or annoying. Annoyance is generally defined as a feeling of displeasure against a source of pollution in the environment which the individual knows or believes will adversely affect his health or well-being (WHO, 1980). As annoyance is a subjective reaction, it has to be evaluated using questionnaire techniques.

When the relationship between noise level and the degree of individual annoyance is evaluated, it is generally found that noise exposure can explain only part of the total reaction. One reason for this is that the noise dose of the individual is not precisely defined in social surveys, because persons living in a specific area are usually allotted the area exposure value. Another reason is the variation in individual sensitivity. The inter-individual variation of an effect, in this case annoyance, caused by an environmental agent is not surprising. Reactions to any type of agent vary among individuals, whether the agent is noise or a chemical substance. The same methodology as is used in toxicology must therefore be applied, i.e. working with average reactions in groups of persons.

A large number of studies has been performed to evaluate the relationship between exposure to road traffic noise and the extent of annoyance in the exposed populations. Based upon methodology originally developed to study the effects of aircraft noise, large investigations in road traffic noise were performed in London by Langdon (1976), followed by a plethora of studies in other countries. It is not within the scope of this book to provide a complete review of all studies performed. Apart from the noise exposure principles, discussed below, there is relatively little controversy regarding the importance of the exposure and the different effects. The studies agree that exposure to road traffic noise is one of the most important sources of annoyance in the general population. The number of persons affected far exceeds the number disturbed by aircraft noise or other environmental noise sources.

As mentioned above, controversy still exists over methods of expressing exposure to road traffic noise. Many studies on road traffic noise demonstrate a fairly linear correlation between the equivalent noise level and the extent of annoyance (Langdon, 1976; Fidell and Baber, 1991) but other studies show a poor correlation (Rylander, 1986). A better relationship was obtained between the extent of annoyance and noise, if the noise levels and number of events were treated separately. An increasing number of events initially caused an increased extent of annoyance, but beyond a saturation point a further increase in the number of events did not influence annoyance. The maximum noise level (MNL) determined the extent of annoyance irrespective of the numbers of events (Rylander *et al.*, 1972; Öhrström *et al.*, 1988).

For continuous noise, there is very little to distinguish between the concepts of equal energy and MNL. This may be one of the reasons why several studies on road traffic noise have been able to show a statistically significant relationship between noise exposure (as average noise levels) and the extent of annoyance. The few studies that have been performed with the aim of investigating the importance of the number of events, the noise levels and the maximum noise levels as independent variables, have mostly come to the conclusion that the relationship between L_{eq} for traffic noise and the effect in the exposed population (whether this be annoyance or sleep disturbance) is less strong than for the MNL concept. As an example, Gjestland (1987) re-evaluated data from a Danish traffic noise study and performed a small laboratory experiment in which he exposed subjects to traffic noise with different levels of heavy traffic. The synthesis from his work suggests that a reduction in the number of heavy vehicles was much more effective than a general noise attenuation. He concluded that the L_{eq}, as a noise index, often fails to describe irregular noise situations.

The noisiest events in road traffic often come from heavy vehicles and there are several reasons why the number of heavy vehicles should be closely related to the extent of annoyance. The noise levels from heavy vehicles are clearly distinguishable from a background of lower noise levels produced by passenger cars. Also, the noise from heavy vehicles has a different acoustic character, and is mostly in the low frequency spectrum. The effect of window attenuation, which is generally poor for low frequencies, results in proportionally higher noise levels indoors from the heavy vehicles. Recent data suggest, however, that light lorries (delivery vans) may be the major source of high noise levels in city traffic. Nevertheless, noisy cars and motorcycles also contribute to the peak levels.

The WHO Task Group on Community Noise noted that noise annoyance varies with activity (Berglund and Lindvall, 1995). The threshold of annoyance for steady-state, continuous noise is around the equivalent sound pressure level of 50 dB(A). Few people are seriously annoyed during the day time at noise levels below about 55 dB(A).

3.5 Conclusions

The range of effects induced by exposure to road traffic noise is wide and covers simple reflex as well as complex psycho-social effects. The most serious adverse effect is sleep disturbance with its long-term consequences for health and well-being. Annoyance due to environmental noise is widespread, particularly in built-up areas and around airports. Annoyance is an important criterion for noise exposure and can be used as the basis for establishing noise control programmes. According to the definition of health given by WHO, subjective annoyance should be considered an important health effect.

Noise standards should relate to the extent of the effect on the population, i.e. what proportion of the population suffering from serious sleep disturbance can be considered as acceptable. According to the principles of risk assessment, the setting of such standards is not in the hands of scientists — it is the responsibility of administrators to choose acceptable levels. Medical effect data constitute the necessary background information for the formulation of these standards. A long-term goal, from a medical point of view, is that the proportion of persons now suffering from sleep disturbance or annoyance needs to be drastically reduced to improve the health of the population.

The following are important research questions that need to be answered in the future:

- Is there a relationship between annoyance and physiological/clinical effects? Some evidence indicates that this is the case for psychiatric symptoms, but more work is required.
- How does interruption of different mental activities relate to physiological/clinical effects and to general annoyance?
- Are activity interference, annoyance reactions and performance effects all secondary reactions to, or symptoms of, noise induced physiological stress?

3.6 References

Abey-Wickrama, I., Brook, M.F. and Gattoni, F.E.G. 1969 Mental hospital admissions and aircraft noise. *Lancet*, **2**, 1275–1277.

Aniansson, G. and Peterson, Y. 1983 Speech intelligibility of normal listeners and persons with impaired hearing in traffic noise. *Journal of Sound Vibration*, **90**, 341–360.

Aniansson, G., Pettersson, K. and Peterson, Y. 1983 Traffic noise annoyance and noise sensitivity in persons with normal and impaired hearing. *Journal of Sound Vibration*, **88**, 85–97.

Babish, W., Ishing, H., Elwood, P.C., Sharp, D.S. and Bainbton, D. 1993 Traffic noise and cardiovascular risk. *Archives Environmental Health*, **48**, 406–413.

Berglund, B. and Lindvall, T., 1995 *Community Noise*. Archives of the Center for Sensory Research, Vol. 2, Issue 1, Stockholm University and Karolinska Institute, Stockholm.

Björkman, M. 1983 Maximum noise level in road traffic noise and cardiovascular risk. *Journal of Sound Vibration*, **127**, 341–360.

Cantrell, W. 1974 Prolonged exposure to intermittent noise, Audiometric biochemical, motor, psychological and sleep effects. *Laryngoscope*, Suppl. 1, **84**, Pt 2.

Cohen, S., Evans, G.W., Krantz, D.S., Stokols, D. and Kelly, K. 1973 Aircraft noise and children, Longitudinal and cross-sectional evidence on adaption to noise and the effectiveness of noise abatement. *Journal of Personality and Social Psychology*, **40**, 330–345.

Dejoy, D.M. 1984 A report on the status of research on the cardiovascular effects of noise. *Noise Control Engineering Journal*, **23**, 32–39.

Fidell, S. and Baber, D.S. 1991 Updating a dosage-effect relationship for the prevalence of annoyance due to general transportation noise. *Journal of the Acoustical Society of America*, **89**, 221–233.

Gattoni, F. and Tarnopolsky, A. 1973 Aircraft noise and psychiatric morbidity. *Psychological Medicine*, **3**, 516–520.

Gjestland, T. 1987 Assessment of annoyance from road traffic noise. *Journal of Sound Vibration*, **112**, 369–375.

ISO 1982 *Acoustics Description and measurement of environmental noise–Part 1: Basic quantities and procedures,* ISO 1996–11982(E) [Replaces ISO R1996, 1971] International Organization for Standardization, Geneva.

Jenkins, L.M., Tarnopolsky, A. and Hand, D. 1979 Comparison of three studies of aircraft noise and psychiatric hospital admissions conducted in the same area. *Psychological Medicine*, **9**, 681–693.

Knipschild, P. and Sallé, H. 1979 Road traffic noise and cardiovascular disease, A population study in the Netherlands. *International Archives of occupational and Environmental Health*, **44**, 55–59.

Langdon, F.J. 1976 Noise nuisance caused by road traffic noise in residential areas. *Journal of Sound Vibration*, **47**, Part I: 243–263, Part II: 265–282.

Öhrström, E. 1989 Sleep disturbance, psycho-social and medical symptoms–a pilot survey among persons exposed to high levels of road traffic noise. *Journal of Sound Vibration*, **133**, 117–128.

Öhrström, E. 1991 Psycho-social effects of traffic noise exposure. *Journal of Sound Vibration*, **151**, 513–517.

Öhrström, E. and Björkman, M. 1983 Sleep disturbance before and after traffic noise attenuation in a block of flats. *Journal of the Acoustical Society of America*, **73**, 887–879.

Öhrström, E., Björkman, M. and Rylander, R. 1988 Noise annoyance with regard to neurophysiological sensitivity, subjective noise sensitivity and personality variables. *Phychol. Med.*, **18**(3), 605–613.

Öhrström, E., Rylander, R. and Björkman, M. 1988 Effects of night time road traffic noise–an overview of laboratory and field studies on noise dose and subjective noise sensitivity. *Journal of Sound Vibration*, **127**, 441–448.

Rylander, R. 1986 Dose-response relationships for traffic noise and annoyance. *Archives of Environmental Health*, **41**, 7–10.

Rylander, R. and Dancer, A. 1978 Startle reactions to simulated sonic boom exposure: Influence of habituation boom level and background noise. *Journal of Sound Vibration*, **61**, 235–243.

Rylander, R., Sörensen, S. and Kajland, A. 1972 Annoyance reactions from aircraft noise exposure. *Journal of Sound Vibration*, **24**, 419–444.

Tarnopolsky, A., Watkins, G. and Hand, D.J. 1980 Aircraft noise and mental health: I Prevalence of individual symptoms. *Psychological Medicine*, **10**, 683–698.

Thachray, R.I. 1972 Sonic boom exposure effects. Startle responses. *Journal of Sound Vibration*, **20**, 519–526.

Vernet, M. 1979 Effect of train noise on sleep for people living in houses bordering the railway line. *Journal of Sound Vibration*, **66**, 483–492.

Wilkinson, R. T. Campbell, K.C. and Roberts, L.D. 1980 Effects of noise at night upon performance during the day. In: *Proceedings of the Third International Congress on Noise as a Public Health Problem*, Freiburg, Germany, ASHA Reports No. 10, 405–412.

WHO 1980 *Noise*. Environmental Health Criteria, 12, World Health Organization, Geneva.

Yiming Z. *et al.* 1991 A dose response relation for noise induced hypertension. *British Journal of Industrial Medicine*, **48**, 179–184.

Chapter 4*

EXPOSURE TO EXHAUST AND EVAPORATIVE EMISSIONS FROM MOTOR VEHICLES

Air pollutants from motor vehicles come from the exhaust pipe, fuel tank, canister and carburettor. Exhaust pipe emissions come from fuel combustion in the engine. Evaporative emissions occur while the vehicle is moving, standing or refuelling. The principal health-related pollutants include CO, NO_X, sulphur oxides, sulphurous and sulphuric acids, reduced sulphur compounds, particulate matter, and aromatic hydrocarbons (HC) such as benzene, toluene, ethyl benzene, the xylenes and the trimethylbenzenes (Atkinson, 1988). Lead (Pb) is a pollutant when it is added to fuel to improve engine performance. Ozone is a secondary air pollutant created in the atmosphere through a photochemical process that involves NO_X, reactive non-methane hydrocarbons and volatile emissions.

Ambient air quality data from fixed-site monitors are typically used to estimate the risk that pollution poses to public health. In general, fixed-site monitors are appropriate in situations where air pollutant concentrations tend to be spatially homogeneous throughout the districts represented by the monitors and where the population of those districts is fairly immobile. Spatial homogeneity also implies that indoor and outdoor pollutant concentrations are nearly the same. Few air pollutants satisfy these assumptions perfectly. The concentrations of many air pollutants from motor vehicles, particularly CO and Pb, are not spatially homogeneous throughout urban atmospheres. For this reason, researchers have used personal exposure monitors to estimate either the total exposure of a population or the exposures of subpopulations in micro-environments that pose higher risks of exposure to these air pollutants. A micro-environment exists for a pollutant if the concentration of the pollutant at a particular location and time is sufficiently homogeneous yet significantly different from the concentrations at other locations (Duan, 1982). Examples of these micro-environments include

* *This chapter was prepared by P.G. Flachsbart*

congested roadways, parking garages, service stations, street canyons and the passenger cabins of motor vehicles in traffic.

Human exposure to an air pollutant occurs whenever a person or population makes contact physically with a pollutant at a particular instant in time (Ott, 1982). An individual's dosage of an air pollutant is affected by the amount of pollutant that enters the body, either through inhalation, ingestion or dermal absorption (Lioy, 1990). Accurate estimation of personal air pollution exposure and dosage are now considered necessary to determine what risk that pollution poses to public health (Sexton and Ryan, 1988).

This chapter selectively reviews studies of human exposure to air pollutants from evaporative fuel losses and exhaust pipe emissions of motor vehicles. The objectives of the chapter are:

- To describe typical levels of human exposure to these air pollutants in various micro environments, and how various factors affect these exposures.
- To estimate how many people are exposed to these pollutants inside vehicles and along roadsides in developed and developing countries.

The first part of the chapter describes typical levels of human exposure to air pollutants from motor vehicles as reported by studies done in various cities world-wide. These levels may not represent current conditions in some cases, because several studies were done prior to the advent of emission controls on automobiles or when many cars lacked modern controls. Personal exposure studies use either a direct or an indirect approach to monitor the air. In the direct approach, the researcher distributes personal exposure monitors (PEMs) to a sample of the population; using PEMs people record exposures to selected air pollutants as they engage in their regular daily activities. In the indirect approach, trained technicians use PEMs to measure pollutant concentrations in selected micro-environments. This information must then be combined with additional data on human activity patterns to estimate the time spent in those micro-environments. Sexton and Ryan (1988) provide further discussion of types of personal monitors and the methodologies used by the direct and indirect approaches.

This chapter does not discuss biological markers to assess exposure, how much time people spend in different micro-environments, or models of exposure. These subjects are discussed by Sexton and Ryan (1988). Recent commuter exposure models have been reviewed by Flachsbart (1993). In addition to these sources, the chapter builds on literature reviews by Sterling and Kobayashi (1977), Flachsbart and Ott (1984), Flachsbart and Ah Yo (1989), Fernandez-Bremauntz (1993) and Ott *et al.* (1994b).

4.1 Exposure assessed using the direct approach

Initially in the direct approach, subjects recorded their exposures and corresponding activities in diaries, as done by a pilot study of CO exposure in Los Angeles, California (Ziskind *et al.*, 1982). This approach was cumbersome and potentially distorted the activity and also later studies applied data loggers to store concentrations electronically, as used by CO exposure studies in Denver, Colorado and Washington, D.C. (Akland *et al.*, 1985). In these studies, subjects still kept diaries to record pertinent information about their activities in specified micro-environments while monitoring personal exposures. All data were then transferred from data loggers and diaries to computer files for analysis.

The direct approach provides a frequency distribution of exposures for a sample of people, selected from either a general or a specific population (defined by demographic, occupational and health-risk factors) for a particular period of interest (e.g. an hour). Studies of either population can then assess what percentage of the population is exposed to pollutant concentrations in excess of ambient air quality standards. Studies of subpopulations (e.g. commuters, taxi and bus drivers, bicyclists, parking garage attendants, etc.) identify those at risk to high exposures in specific micro-environments. Examples of each type of study are discussed below.

4.1.1 General population studies

Direct studies of the general population are rare because of their expense and the logistical problems of monitor distribution. The best examples are the Denver, Colorado and Washington, D.C. studies of CO exposures (Akland *et al.*, 1985). In both studies, the target population included non-institutionalised, non-smoking residents, who were aged 18–70 and lived in the city's metropolitan area during the winter of 1982–83. Both studies found higher exposures associated with commuting. The two highest average CO concentrations occurred when subjects were in a parking garage or parking lot and when travelling. High exposures were also traced to indoor and occupational sources. The average CO concentrations observed for all "in-transit" micro environments are shown in Table 4.1, based on supplementary data analyses by Ott *et al.* (1994b). Denver had higher averages for each micro-environment, because its colder climate contributed to more CO emissions from motor vehicles. Both studies revealed the extent of the disparity between estimates of CO exposure based on fixed-site and personal monitors. In both cities, the composite network of fixed-site monitors overestimated the

Table 4.1 Typical in-vehicle CO exposures in two cities in the USA, 1982–83

In-transit micro-environment	Denver, Colorado			Washington, D.C.		
	n	Mean (ppm)	s.d.	n	Mean (ppm)	s.d.
Motorcars	3,029	7.8	11.3	3,345	4.6	6.5
Buses	70	9.0	6.9	167	3.5	5.9
Lorries	350	7.2	9.5	135	5.3	8.1
Motorcycles	20	12.3	9.5	3	3.0	3.3
Subways	–	–	–	95	2.0	1.8

– No data available
s.d. Standard deviation
n Number of observations
Source: Ott *et al.*, 1994b

8-hour personal exposures of people with low level exposures and underestimated the 8-hour personal exposures of people with high exposures. In Denver, over 10 per cent of the daily maximum 8-hour personal exposures exceeded 9 ppm, the U.S. National Ambient Air Quality Standard (NAAQS). In Washington, D.C., about 4 per cent of maximum 8-hour personal exposures exceeded 9 ppm.

4.1.2 Specific population studies

Commuter exposure

Cortese and Spengler (1976) found that CO exposures averaged 11.9 ppm for 66 non-smoking volunteers who commuted 45–60 minutes one way in Boston, Massachusetts. Each volunteer wore a CO personal monitor for 3–5 days over a six-month period between October, 1974 and February, 1975. Simultaneous measurements taken at six fixed-site monitors averaged only 6 ppm. Motorcar commuters had exposures nearly twice that of transit users, and about 1.6 times that of split-mode commuting. Shikiya *et al.* (1989) collected samples of in-vehicle concentrations of CO, two aldehydes, six halogenated hydrocarbons and four metals during the peak commuting hours during summer and winter in metropolitan Los Angeles, California. The researchers selected a random sample of 140 non-smokers who commuted from home to work in privately owned vehicles during each season. Driving patterns and ventilation conditions were not controlled. Table 4.2 gives the average concentrations of organic gases and metals for round-trip commuting journeys. The one-way commuting collection time averaged 33 minutes for organic gases and aldehydes and 52 minutes for metals. Mean in-vehicle

Table 4.2 Typical Concentrations in Los Angeles, California, in 1987–88

Air pollutant	Mean in-vehicle concentration	Mean ambient concentration
Organic gases (ppb)		
Benzene	13.3	5.3
Carbon monoxide	8,599	3,661
Ethylene dibromide	0.014	0.016
Ethylene dichloride	0.033	0.010
Formaldehyde	12.5	6.8
Toluene	36.3	14.7
Xylene	32.9	15.3
Metals ($\mu g\ m^{-3}$)		
Chromium	0.012	0.023
Lead	0.218	0.208

Source: Shikiya *et al.*, 1989

Table 4.3 Typical 8-hour CO exposures in Denver, Colorado, 1978–79

Type of exposure	Non smokers		Smokers	
	Median (ppm)	Range (ppm)	Median (ppm)	Range (ppm)
Office	7.5	4.2–22.1	7.1	4.1–9.6
Traffic	20.4	7.8–44.3	22.1	8.0–55.3

Source: Jabara *et al.*, 1980

concentrations substantially exceeded mean ambient concentrations for all pollutants except ethylene dibromide, chromium and Pb.

Occupational exposure

Jabara *et al.* (1980) collected 8-hour CO exposures for 65 employees (including smokers and non-smokers) of the Denver, Colorado Police Department, who worked near heavy traffic for extended time periods. A group of 33 office employees served as controls for comparison. Table 4.3 gives the 8-hour CO concentrations of each type of employee. The higher median concentration for office non-smokers was attributed to five high readings among employees who worked in offices above a parking garage in the police station. The study concluded that CO leaked upstairs from the garage on very cold mornings when garage doors were closed and vehicles engines were running to warm up before leaving.

Nagda and Koontz (1985) reported the CO exposures of a sample of 58 people divided into three subgroups (housewives, office employees and construction workers). Each person wore a personal monitor for four consecutive 24-hour periods in Washington, D.C. during the autumn of 1982. The mean exposure of each group was similar, but slightly higher for the office and construction workers who spent relatively more time in travel activities. Higher exposures occurred during rush hours and in heavy traffic.

Subida and Torres (1991) determined the exposures of jeepney drivers to air pollutants along 24 routes in metropolitan Manila, Philippines between May, 1990 and May, 1991. Of 72 jeepney drivers monitored, 93 per cent had 24-hour cumulative exposures to matter SPM that were 2–10 times higher than the guidelines of the World Health Organization (WHO, 1987), and 100 per cent were exposed to Pb beyond WHO's annual average guideline of 0.5–1.0 $\mu g\ m^{-3}$. Jeepney drivers, who worked about 10–12 hours a day also had mean 24-hour exposures to CO and SO_2 that were above WHO's guidelines. For comparison, the mean 24-hour exposures to SPM and Pb of air-conditioned bus drivers and office building commuters were found to be relatively lower but still above WHO's guidelines in some cases. In the Philippines, 28 per cent of all vehicles use diesel fuel, and the Pb content in petrol is high as it ranging from 0.6 to 0.8 $g\ l^{-1}$.

4.2 Exposure assessed using the indirect approach

Exposure studies that have used the indirect approach are more numerous than those that have used the direct approach. Studies of the exposures of travellers discussed first; where some studies have concentrated on travel over selected routes in urban areas and collected data at standardised intervals in time and space, using either real or hypothetical commuters. Other studies have focused solely on certain types of travel (e.g. work trips), especially where vehicles move slowly due to high traffic volumes and/or constricted space. More elaborate studies have involved many routes, multiple modes of travel, and two or more times of the day and seasons of the year. Some studies have also tested vehicles under different ventilation conditions.

Many of the remaining studies that have used the indirect approach measured exposures in commercial districts and public facilities subject to motor vehicle and pedestrian traffic. These studies focused on indoor settings, such as retail stores and office buildings on busy streets, and on outdoor settings, such as sidewalks, street intersections, and service stations. Other studies focused on confined spaces used by motor vehicles, such as parking garages, tunnels and underpasses. A few of these studies measured air quality inside shopping centres and in office buildings where pollution could be traced to attached parking garages.

4.2.1 Exposure by travellers

Early studies of actual travellers associated their exposures with defective exhaust systems and cars without effective emission controls. Later studies of real and hypothetical travellers found that their exposures varied according to the condition of the roadway and its functional type, by the time of day and season of year when travel occurred, by the type of vehicle used, and by differences in vehicular ventilation. These studies have also shown that in-vehicle exposures differed from ambient concentrations measured at fixed-site monitors, that trends in exposures have occurred over time, and that the exposures of those in developed countries differ from those in developing countries.

Exhaust systems and emission controls

Amiro (1969) found that engine and/or exhaust pipe emissions of CO leaked into the passenger cabins of 9 out of the 19 vehicles studied. Concentrations of CO inside contaminated vehicles ranged up to 400 ppm. Clements (1978) reported that of 645 school buses tested, 7.2 per cent had average readings in excess of 20 ppm and 5.4 per cent had maximum readings above 50 ppm. He estimated that CO concentrations in excess of 20 ppm existed in school buses used by up to 2.1 million school children in the USA on a daily basis. Ziskind *et al.* (1981) identified which sustained-use vehicles (buses, taxis and police cars) had faulty exhaust systems, i.e. that leaked CO into passenger cabins. They found that 58 per cent of 120 rides taken in faulty vehicles exceeded the 8-hour NAAQS of 9 ppm.

Chaney (1978) demonstrated the effectiveness of emission controls in the USA in reducing personal CO exposure. He assumed that the year of manufacture of a vehicle could be used as an indication of its levels of emission control, because the USA adopted nationwide emission controls for automobiles in 1968. During a cross-country trip, he showed that a greater percentage of the total increase in concentration inside his vehicle could be attributed to a small percentage of older cars. Only 3.3 per cent of motor vehicles that passed his car were made before 1970, but collectively they were responsible for 45 per cent of the total increase in CO concentration inside his car. Conversely, 76 per cent of all vehicles that passed his car were made between 1970 and 1977, but together they contributed only 12 per cent of the total increase in CO inside his car.

Roadway location and function

The location of a roadway and its functional type can act as surrogate measures for several factors. The speed of traffic affects vehicular emission rates and traffic volume affects total emissions. Traffic speed and volume are interrelated and are themselves affected by the capacity of the roadway. Also, a roadway's degree of enclosure affects concentrations of air pollutants emitted by vehicles. Two early studies showed the effects of these factors on pollutant exposure. Brice and Roesler (1966) reported that CO and HC exposures inside a moving vehicle were higher on streets in downtown Cincinnati, Ohio than on major arteries and expressways, because tall buildings reduced ventilation. Similarly, Lynn *et al.* (1967) observed that average CO exposures for 30-minute drives in 14 USA cities were highest in central cities (31.9 ppm), than in arterials (24.6 ppm) and expressways (19.4 ppm).

Some studies determined CO exposures for different traffic volumes that were rated subjectively. Godin *et al.* (1972) found that in-vehicle CO exposures in Toronto, Canada were 20 ppm higher in "heavy" traffic than in "light" traffic as subjectively rated on an ordinal scale. Ten years later, Petersen and Allen (1982) reported that in-vehicle CO concentrations were only 4.3–6.3 ppm higher in "heavy" traffic than in "light" traffic during 3-hour drives in Los Angeles, California.

Studies of bicyclists also show the effects of different types of roadways on exposure. In Boston, Massachusetts, Kleiner and Spengler (1976) found that the mean CO exposures of bicyclists were highest for "narrow, busy two-lane" streets, intermediate for "wide, busy, four-lane" streets, and lowest for "narrow, light, two-lane" streets regardless of travel period. In Southampton, England, Bevan *et al.* (1991) found that the mean CO exposures of bicycle commuters were 13 ppm on an urban route subject to a high level of traffic, but only 7.9 ppm on a suburban route that had relatively less traffic. They also found similar results for respirable SPM and many aromatic VOCs.

Passengers may be exposed to extremely high pollutant levels when motor vehicles are idling in a queue. Myronuk (1977) measured CO concentrations inside an automobile at space-confined, drive-up facilities in Santa Clara Valley, California. The in-vehicle CO concentrations ranged from 15–95 ppm for 15-minute averages, with short-term peaks between 100 and 1,000 ppm. Background CO concentrations were only 2–5 ppm. Monitoring occurred when winds were low.

More recent studies attempted to quantify the effect of traffic volume and speed on exposure. Flachsbart *et al.* (1987) reported that in-vehicle CO exposures fell by 35 per cent when the speed of test vehicles increased from 10 to 60 miles per hour on eight commuter routes in Washington, D.C. Koushki *et al.* (1992) confirmed this finding in a similar study of a typical commuter route in central Riyadh, Saudi Arabia. They also found that an increase in traffic volumes from 1,000 to 5,000 vehicles per hour corresponded to a 71.5 per cent increase in mean in-vehicle CO concentrations.

Researchers often assume that in-vehicle exposure does not vary greatly from lane to lane on a given highway. Flachsbart (1989) found that priority (with-flow and contra-flow) lanes were effective in reducing exposure to motor vehicle exhaust on a coastal artery in Honolulu, Hawaii. Compared with commuter CO exposure in adjacent but congested lanes, exposure in priority lanes was about 18 per cent less for those in shared cars, 28 per cent less for those in high-occupancy vehicles (e.g. shared vans), and 61 per cent less for those in express buses. These differences occurred possibly because commuters in priority lanes travelled faster than those in the congested lanes. The motion of faster vehicles created more turbulence which may have helped to disperse pollutants surrounding vehicles in priority lanes. Furthermore, these differences existed even though the priority lanes were often downwind of the congested lanes. Although higher speeds were related to lower exposures in priority lanes, differences in exposure could also have been due to differences in vehicle type and ventilation, which were not controlled.

Recent studies have looked at the effect of routes in different locations on exposure to CO and other air pollutants. Chan *et al.* (1991a) reported significantly different in-vehicle exposures to CO and VOCs, but not to NO_2 and O_3, for standardised drives on three routes that varied in traffic volume and speed in Raleigh, North Carolina. The highest CO and VOC concentrations occurred in the downtown area, which had heavy traffic volumes, slow speeds and frequent stops. The next highest concentrations occurred on an interstate beltway (highway ring road), which had moderate traffic volumes and high speeds, and the lowest concentrations occurred on rural highways, which had low traffic volumes and moderate speeds. Chan *et al.* (1991b) reported a similar result in Boston, Massachusetts, where in-vehicle concentrations of benzene, toluene, ethyl benzene, m-/p-xylene, and o-xylene were 1.5 times greater for commuters on urban routes than for commuters on interstate routes. Likewise, Dor *et al.* (1995) reported higher exposures to CO and six monocyclic aromatic hydrocarbons (MAHs) for a route through central Paris than for two suburban routes.

Diurnal and seasonal variation

Haagen-Smit (1966) found preliminary evidence in the 1960s that CO exposures during afternoon commuting were greater than during morning commuting in Los Angeles, California. Similar results were found by Cortese and Spengler (1976) in Boston, Massachusetts, by Wallace (1979) in Washington, D.C., and by Dor *et al.* (1995) in Paris, France. However, Holland (1983) found contrary evidence in four USA cities. In each study, exposure differences by time of day were not statistically significant. Variations in traffic volumes and speeds, ambient concentrations, and/or meteorological conditions during different periods of the day could have been offsetting effects on exposure.

Seasonal variations in ambient temperatures and traffic volumes affect total automotive emissions. Likewise, seasonal variations in wind speeds affect the dilution and dispersion of those emissions on roadways. Ott *et al.* (1994a) in northern California and Dor *et al.* (1995) in France, who both measured in-vehicle CO exposures for an entire year, reported that they were generally higher in autumn/winter than in spring/summer. This result was attributed to colder winter temperatures which increase CO emissions per vehicle mile. Higher summer temperatures have the opposite effect.

Seasonal changes may have a different effect on exposure to VOCs. Weisel *et al.* (1992) reported that average in-vehicle exposures to benzene, toluene, m-xylene, p-xylene and o-xylene after a 30-minute engine idling in New Brunswick, New Jersey, were greater during summer than winter. They concluded that more evaporation of these fuel components from the fuel tank and engine may occur during summer, despite seasonal adjustments made to the Reid vapour pressure (RVP) of petrol to counter this effect. In a direct study of Los Angeles commuters, Shikiya et al. (1989) reported a result for CO which was similar to findings by Ott *et al.* (1994a) and Dor *et al.* (1995), but results contrary to Weisel *et al.* (1992) for benzene and toluene. However, contrary results could be due to regional differences in meteorological conditions which affect ambient VOC concentrations.

Mode of travel

Flachsbart *et al.* (1987) reported that in-vehicle CO exposure varied substantially by travel mode for hypothetical commuters in the Washington, D.C., metropolitan area during winter 1983. The time-weighted average CO concentrations ranged from 9–14 ppm for automobile trips of 40–60 minutes on eight congested highway routes. By comparison, typical CO exposures for mass transit users were substantially less. They ranged from 4–8 ppm for bus trips of 90–110 minutes on four routes, and from 2–5 ppm for rail trips of 30–45 minutes on three routes. Joumard (1991) and Dor *et al.* (1995) found similar differences in CO exposures for public and private modes of travel in French cities and towns.

Table 4.4 Typical CO concentrations in Mexico City, Mexico, 1991

Travel mode	Range of mean CO concentrations (ppm)	Number of trips
Private cars	55–57	34
"Combi" type vans	39–67	35
Minibuses	32–64	152
Diesel buses	20–40	170
Electric trolleybuses	22–32	47
Subways and light rails	16–26	111

Source: Fernandez-Bremauntz, 1993

Flachsbart *et al.* (1987) observed higher CO exposures for automobile commuters than did Akland *et al.* (1985), even though both studies occurred in Washington, D.C., at about the same time. This difference occurred because Akland *et al.* (1985) used the direct approach and Flachsbart *et al.* (1987) used the indirect approach. Akland *et al.* (1985) directly surveyed a representative sample, thereby capturing all types of urban travel throughout the day, while Flachsbart *et al.* (1987) selected long commuter routes over known congested highways during rush-hour periods only. Thus, the results from Flachsbart *et al.* (1987) may be representative of people with that type of commuting pattern. In a recent study, Fernandez-Bremauntz (1993) monitored CO concentrations inside six types of vehicles typically used for commuting in the metropolitan area of Mexico City, Mexico, during winter 1991. Table 4.4 reports results for 549 trips, which represented a combination of five different corridors during morning and evening commuting periods. Generally much higher exposures were found inside automobiles than inside buses and subways, a result similar to the findings discovered earlier for Washington, D.C. and Paris. It was concluded that this variation could be caused by variation in the height of the personal exposure monitor above the roadway and by differences in the ventilation system of each mode of transport.

Ventilation

Petersen and Sabersky (1975) discovered that the passenger cabins of motor vehicles have a dampening effect on exterior CO concentrations. Concentrations measured outside vehicles exhibited rapid variations and high peaks, while interior concentrations showed more gradual variations. Similar findings were made by Colwill and Hickman (1980) and Koushki *et al.* (1992). Chaney (1978) found that the rapid fluctuations in exterior concentrations reflected the exhaust from passing vehicles.

Three studies have measured interior and exterior CO concentrations simultaneously during trips. The mean ratio of interior to exterior concentrations was similar for two of the studies. Petersen and Allen (1982) reported a ratio of 0.92, while Koushki *et al.* (1992) reported 0.84. Petersen and Allen (1982) reported that their ratio remained about the same when windows and vents were closed and when the air conditioner was in use. Colwill and Hickman (1980) reported that internal CO concentrations were about 30–80 per cent of exterior concentrations, but this result could be explained by the different location of their exterior monitor which was at bumper height. The other two studies placed the exterior monitor higher on the car.

Flachsbart *et al.* (1987) found that a highly polluted atmosphere inside a garage could increase the CO concentration inside a vehicle while parked in the garage. Once the vehicle was driven away from the garage, the high concentration eventually dissipated, but only slowly because the car windows were usually kept closed.

Comparisons with fixed-site monitors

Many studies have shown that fixed-site monitors have underestimated in-vehicle exposures to CO and several VOCs, but have overestimated exposures to NO, NO_X and O_3. A summary of a few of these studies is given below.

Petersen and Sabersky (1975) reported that average in-vehicle CO concentrations in Los Angeles, California were 15–20 ppm, but fixed-site monitors showed maximum CO concentrations of only 8 ppm. In a later study of Los Angeles, Petersen and Allen (1982) reported lower in-vehicle CO values of 10.9–15.3 ppm, but these values were still higher than concentrations at fixed-site monitors by a factor of 3.9. Holland (1983) found that fixed monitoring stations in residential areas of four USA cities underestimated the time-weighted mean of commuting and residential driving exposures to CO over hypothetical routes by factors of 0.4–0.7. In Raleigh, North Carolina, Chan *et al.* (1991a) reported that median CO concentrations were 11 ppm inside test vehicles on hypothetical routes, but only 2.8 ppm at fixed-site monitors. In a review of 16 USA studies carried out between 1965 and 1992, Flachsbart (1995) reported that mean CO exposure was typically 3.5 times greater than the mean ambient CO concentration. Somewhat higher ratios of commuter to ambient concentrations were observed by studies outside the USA. Tonkelar (1983) reported a mean CO concentration of 5.8 ppm for 4,231 trips at peak hours in Delft, The Netherlands, which was five times the concurrent CO concentration observed at fixed-site monitors. Fernandez-Bremauntz (1993) found that CO exposures in automobiles exceeded ambient concentrations by factors ranging from 3.5–6.9 in Mexico City, Mexico.

Two recent studies agree that in-vehicle exposures to VOCs are higher than their ambient concentrations, but disagree on the ratio of personal to fixed-site concentrations. Chan *et al.* (1991a) reported that the median VOC exposures inside two test vehicles in Raleigh, North Carolina, were from 5–10 times higher than the median ambient concentrations at fixed-site monitors for benzene, ethyl benzene, hexane, isopentane, m-/p-xylene, n-butane, n-pentane, o-xylene, 1,2,4-trimethylbenzene, toluene, 2-methyl pentane, 2,3,4-trimethylpentane and 2,2,4-trimethylpentane. Weisel *et al.* (1992) found that mean in-vehicle concentrations of benzene, m-/p-xylene, and o-xylene were 3–12 times higher than ambient levels during community journeys between New Brunswick, New Jersey and New York City.

Studies show that fixed-site monitors overestimated in-vehicle exposures to other primary and secondary air pollutants. Petersen and Sabersky (1975) found that in-vehicle concentrations in Los Angeles, California, ranged from 0.05–0.50 ppm for NO and from 0.24–0.80 ppm for NO_X. Ambient concentrations of these two pollutants were significantly higher at the nearest fixed-site stations. Petersen and Sabersky (1975) also found that O_3 concentrations were generally below 0.05 ppm inside their vehicle, which was three to four times lower than O_3 levels in the ambient environment. Chan *et al.* (1991a) observed a similar result for O_3 in Raleigh, North Carolina.

Trends in exposure

Several studies indicate trends in commuter exposure to CO and Pb in the USA. Ott *et al.* (1994a) measured in-vehicle CO concentrations on 88 standardised trips over a one year period in 1980-81 on a suburban highway near San Jose, California. They reported a mean CO concentration of 9.8 ppm for trips of 35–45 minutes. In 1991–92, Ott *et al.* (1993) surveyed this highway again using a methodology similar to their previous study to determine in-vehicle exposure trends. They reported that the mean in-vehicle CO concentration had dropped to 4.6 ppm or 46.9 per cent of the mean value estimated 11 years earlier. They attributed the exposure reduction to replacement of older vehicles with newer ones that have lower CO emission factors. This reduction is particularly significant, as daily traffic volumes on this highway grew by 19.1 per cent during the intervening period, according to estimates by Yu *et al.* (1994).

Flachsbart (1995) has reported a long-term downward trend for in-vehicle exposure to CO in traffic, based on a review of 16 USA studies published between 1965 and 1992. He concluded that emission controls on motor vehicles sold in the USA had reduced in-vehicle CO exposure almost 92 per cent over this period, despite dramatic growth in the motor vehicle population and

Table 4.5 Mean in-vehicle CO concentrations for different modes of travel in three cities in different countries

	CO concentration (ppm)		
Travel mode	Washington, USA 1983	Riyadh, Saudi Arabia 1986–87	Mexico City, Mexico 1991
Motor cars	9–14	29.5[1]	55–57
Diesel buses	4–8		20–40
Rail transits	2–5		16–26

[1] Non-smoking cases only

Sources: Flachsbart *et al.*, 1987; Koushki *et al.*, 1992; Fernandez-Bremauntz, 1993

total miles travelled. This implies that levels of in-vehicle CO exposure reported in the past for a particular place and time in the USA may not be indicative of exposures in that same place at the current time.

Reductions in the Pb exposure of USA commuters are evident from two studies done a decade apart. Dzubay et al. (1979), who used the indirect approach, reported that the mean in-vehicle Pb concentration on Los Angeles freeways was 10.9 $\mu g\ m^{-3}$ for drives of 2–4 hours. This Pb exposure was six times greater than the mean ambient concentration. Ten years later, Shikiya *et al.* (1989), who used the direct approach, reported a mean in-vehicle Pb concentration of 0.218 $\mu g\ m^{-3}$. This indicates a 98 per cent reduction in the Pb exposure of commuters, which was achieved by phased reductions of Pb from petrol, as mandated by the US EPA during the period between the studies.

Country comparisons

Studies by Flachsbart *et al.* (1987), Koushki *et al.* (1992) and Fernandez-Bremauntz (1993) were carried out in the USA, Saudi Arabia and Mexico, respectively. These studies used similar methods of data collection and analysis, with one exception. Smoking sometimes occurred in vehicles used in Saudi Arabia, but not in vehicles used in the USA and Mexico. Table 4.5 shows that the mean in-vehicle CO exposures for automobile, diesel bus and rail transit in Mexico City were much higher than for comparable modes in Washington, D.C. The typical exposure of non-smoking, automobile commuters in Riyadh, Saudi Arabia, was at an intermediate level. This comparison illustrates the effectiveness of automotive emission controls in reducing exposure. The USA initiated nationwide emission controls on automobiles in 1968, and adopted progressively tighter controls thereafter. Similar controls did not exist in Mexico and Saudi Arabia when the commuter exposure studies were performed in those countries. The exposures reported recently for Mexico are even higher than exposures observed in the USA prior to its adoption of emission controls. Fernandez-Bremauntz (1993) reported mean CO exposures of 55–57 ppm for automobile trips in Mexico City, while Brice and Roesler (1966) reported mean CO exposures of 21–40 ppm for automobile trips in five USA cities (Chicago, Cincinnati, Denver, St. Louis and Washington, D.C.). Higher exposures in Mexico City could also be due to the city's larger size and unfavourable topography; it had 2.5 million motor vehicles with largely uncontrolled emissions in 1989, and it sits in a basin surrounded by mountains (UNEP/WHO, 1994).

Not all developing countries have high CO exposures in urban areas. Chan *et al.* (1994) reported that the average CO exposure of bicycle commuters ranged from 3.7–8.2 ppm along four bus routes in Guangzhou, China. Exposures did not exceed the Chinese national assessment standard. A few years earlier, Bevan *et al.* (1991) reported CO exposures for bicyclists in England that were higher than those reported by Chan *et al.* (1994) in China. In Southampton, bicycle commuters had mean CO exposures of 10.5 ppm for 16 trips that lasted 36 minutes on average. Trips were evenly distributed between an urban route and a suburban route and between morning and evening rush-hour periods. It can be speculated as to why exposures were higher in England than in China, even though motor vehicles in England have emission controls, and Chinese cars do not, Southampton probably has a higher concentration of vehicles on its roadways than does Guangzhou. Hence, fewer automobiles on roads in China may help offset its lack of emission controls on cars.

By contrast commuters in developing countries may have higher VOC exposures. Chan *et al.* (1993) reported that students who commuted by motorcycle in Taipei City, Taiwan, were exposed to VOC concentrations that were three to eight times higher than those reported by Shikiya *et al.* (1989) for automobile commuters in Los Angeles, California. In Taipei City, the mean benzene exposure of motorcycle commuters was 379.7 $\mu g\ m^{-3}$ for trips lasting 17 minutes on average. In Los Angeles, the benzene exposures of automobile commuters ranged from 31.2–50.4 $\mu g\ m^{-3}$ for trips with an average duration of 33 minutes.

4.2.2 Commercial districts and public facilities

The commercial districts and public facilities of urban areas attract and generate relatively large volumes of traffic that typically circulate at low speeds with frequent stops and starts. This traffic pattern produces relatively high CO emissions. Hot exhaust gases rise from exhaust pipes and are dispersed by wind. If emissions occur in open areas, they may be dispersed by wind and concentrations diminish with greater distance downwind of the source. If emissions occur in street canyons, high concentrations of pollutants may result at street level. The potential for human exposure is great because commercial districts and public facilities attract large numbers of people. The first two sections below discuss spatial and temporal variations in exposure and indoor-outdoor relationships. The last two sections compare exposures in different countries and measurements from personal and fixed-site monitors.

Spatial and temporal variations

An early study in the USA explored the spatial variation in outdoor pollutant concentrations. Colucci and Begeman (1969) found that outdoor mean CO concentrations were usually the highest, but varied the most, in commercial districts (3.5–10 ppm) of Detroit, New York and Los Angeles. By comparison, CO concentrations were lower near freeways (6–8 ppm) and lowest in residential areas (2.5–5.5 ppm). They also found that personal CO exposures in New York and Los Angeles tended to be higher during summer and autumn when average wind speeds were generally lower.

Later studies looked at how exposures varied with distance from the source. Besner and Atkins (1970) reported that CO concentrations declined with greater distance from an expressway in an open area of Austin, Texas. At 16 feet from the road, CO concentrations ranged from 3.4–6.0 ppm, while at 95 feet concentrations ranged from 2.4–3.9 ppm. Rashidi and Massoudi (1980) studied eight shallow street canyons in central Tehran, Iran. They found that sidewalk concentrations of CO measured 50 m from the street centre line were only 85 per cent of the centre line concentrations. Sidewalk concentrations were high when traffic flow was high and moving at moderate speed; but concentrations were also high when flows were lower and vehicles were moving at very low speeds due to congestion. The study concluded that traffic flow within a certain range is more important than vehicle speed in generating high CO emissions.

Spatial and temporal variations have also been found in Canada. In Toronto, Godin *et al.* (1972) observed that CO concentrations were highest in the downtown area, intermediate at a suburban home, and lowest at a semi-rural farm house. Another Canadian study showed how concentrations can be reduced by closing a street and traffic creating a pedestrian mall. Wright *et al.* (1975) measured 4–6 minutes average CO exposures of pedestrians and street-workers during the summer and autumn of 1973 in Toronto. Before the street was closed, the average CO concentrations at two street junctions were

9.4 ± 4.0 ppm (one standard deviation) and 7.9 ± 1.9 ppm. After the street was closed, the averages dropped to 3.7 ± 0.5 ppm and 4.0 ± 1.0 ppm, respectively, roughly equivalent to the urban background level.

Ott and Flachsbart (1982) observed that CO concentrations in 55 out of 74 commercial settings in five California cities were stable for nearly two hours. This indicated that the exposure associated with a short visit (e.g. 3 minutes or less) to a particular setting could be representative of a longer period. They found that CO levels in outdoor settings tended to vary more rapidly with time than in indoor settings. Compared with momentary fluctuations in CO concentrations at a given setting, concentrations differed more from one setting to another, and even more from one date to another for the same setting. Daily variation in CO concentrations occurred partly because of variation in wind speeds. Lower CO values were encountered on windier days.

Sisovic and Fugas (1987) observed a seasonal difference in CO exposures for a commercial district in Zagreb, Yugoslavia (now Croatia). Walking tours of busy narrow streets occurred between 0900h and 1400h and lasted between 160 and 195 minutes. During the walking tour, the mean CO concentration was higher in winter (5.9 ppm) than in summer (3.0 ppm). Average CO concentrations inside six shops visited during the tour ranged from 7.6 to 13.9 ppm during winter and from 5.9 to 14.0 ppm during summer. Higher average concentrations occurred in five shops during the winter, and these increases were attributed to space heating and greater vehicular emissions in cold weather. The proximity and density of traffic explained differences in CO levels between sites.

More recently, Luria *et al.* (1990) found that the combination of roadway design and traffic volume had a significant effect on road-centre concentrations of CO and NO_2 in Jerusalem, Israel, and that macro meteorological conditions had no effect. Two sections of a 6 km route each had high but similar concentrations of these pollutants, but differed substantially in roadway geometry, land use and traffic volume. The narrower section, which was confined by continuous rows of commercial buildings, attracted 15,000 vehicles per 12-hour period and many pedestrians. The wider section which had fewer buildings and pedestrians, carried 36,000 vehicles per 12-hour period. In effect, average road-centre concentrations in a street canyon with less traffic volume were roughly equivalent to concentrations in a wider, more heavily travelled road. Hence, a roadway's cross section, and particularly its degree of confinement, may have a greater effect on personal exposure than a roadway's traffic volume. The speed of traffic on each roadway section was not reported.

Indoor-outdoor relationships

Most studies of indoor exposure invariably also measure outdoor concentrations to determine their relationship. Yocum *et al.* (1971) found that indoor CO concentrations in Hartford, Connecticut, were similar to outdoor concentrations, but lagged behind the peak concentrations observed outdoors. Godin *et al.* (1972) reported similar findings for downtown offices in Toronto, Canada. They found that indoor CO concentrations matched outdoor levels, but with a lag of 1–2 hours.

The General Electric Company (1972) measured CO concentrations in two high-rise buildings in New York City. One building had been built over a highway. They found that indoor concentrations normally were lower than outdoor concentrations at all heights above the roadway when outdoor concentrations were high. Conversely, when the outdoor concentrations were low, indoor concentrations were not as low. In the absence of indoor CO sources, indoor concentrations followed outdoor levels with some degree of time lag, but with a tendency not to reach either the extreme high or low values that were found outdoors. The study also reported that at heights greater than 30 metres above the roadway, CO concentrations were greater indoors than outdoors. This result was attributed to the trapping of CO within the building.

Bellin and Spengler (1980) reported mean CO concentrations for 15-minute periods at a terminal of Boston's Logan International Airport. The indoor (ticket counter) concentrations were significantly lower than the outdoor (curb-side) concentrations. The 95 per cent confidence interval for all indoor locations (5.7–6.8 ppm) was lower than the interval for all outdoor locations (9.7–12.5 ppm). The 1-hour maximum concentrations were 15 ppm (indoors) and 23 ppm (outdoors), both occurring on the same day.

Ott and Flachsbart (1982) measured indoor and outdoor CO concentrations for 88 sites in the Union Square district of San Francisco, California. They reported that indoor CO concentrations were statistically, but not substantially, less than those outdoors. At each site, the entrance to the building was closed, and both indoor and outdoor measurements were taken less than 6m apart and within 3 minutes of each other.

Country comparisons

Two studies by the same research team showed that sidewalk CO exposures were similar in two USA cities. Wilson and Schweiss (1978b) first measured CO concentrations at 36 outdoor sites (including small parks and isolated areas) and two sidewalks in downtown Seattle, Washington. The 4-hour average CO concentrations along sidewalks ranged from 1.1 to 11.9 ppm. Wilson and Schweiss (1978a) then measured CO concentrations at 40 outdoor sites and four sidewalks in Boise, Idaho. Pedestrian exposures ranged from 2.9 to 14.0 ppm for 2–4 hour averages. The highest 8-hour CO average was 17.2 ppm.

A few years later, Ott and Flachsbart (1982) took instantaneous CO concentrations at 1-minute intervals while walking on sidewalks of busy streets and into nearby stores, hotels, office buildings, restaurants, etc. located in five California cities. Concentrations of CO were above zero in most indoor and outdoor settings due to emissions from nearby traffic, but rarely were they higher than the USA. ambient air quality standards of 9 ppm for 8 hours and 35 ppm for 1 hour.

Studies in Iran and Thailand revealed higher curb-side concentrations than in the USA. Rashidi and Massoudi (1980) reported that CO concentrations ranged up to 140 ppm for 5-minute averages on sidewalks in Tehran, Iran. In Bangkok, Thailand, mean CO concentrations for 17 curb-side locations ranged from 3 to 24 mg m^{-3} for 1 hour, and from 2 to 26 mg m^{-3} for 8 hours. In Thailand, the 1-hour and 8-hour CO standards are 50 and 20 mg m^{-3}, respectively. Furthermore, daily concentrations of particulate matter slightly exceeded the ambient standard of 330 μg m^{-3}, and at one location these concentrations were two to three times higher than the standard. Average daily Pb concentrations at 18 sites monitored ranged from 0.6 to 3.5 μg m^{-3}, which was below the ambient standard of 10 μg m^{-3} (Office of the National Environment Board, 1989).

Comparisons with fixed-site monitors

Studies of downtown areas were among the first to show that CO concentrations along sidewalks were greater than ambient levels recorded at fixed-site monitors. Ott and Eliassen (1973) reported average CO concentrations in the range 5.2–14.2 ppm for sidewalks along congested streets of downtown San Jose, California. Corresponding CO averages at fixed-site monitors ranged from 2.4–6.2 ppm. A decade later, Ott and Flachsbart (1982) found a narrower gap between simultaneous CO measurements from fixed-site monitors (FSMs) and PEMs deployed at indoor and outdoor commercial settings in five California cities. The mean CO concentrations were 2.0 ppm (FSM), 3.0 ppm (PEM-indoor) and 4.0 ppm (PEM-outdoor). More recently, Chan *et al.* (1991a) found ratios of about 5:2 between sidewalk and fixed-site median concentrations of benzene, toluene and m-/p-xylene in Raleigh, North Carolina.

4.2.3 Street intersections

Emissions of CO and roadside concentrations can increase dramatically whenever petrol driven vehicles stop and idle at street intersections or otherwise form a queue. The severity of exposure partly depends on how much traffic is handled by the intersection and the distance of an individual from it.

Ramsey (1966) surveyed 50 intersections over a 6-month period in Dayton, Ohio. He observed a strong relationship between traffic volume and ambient CO concentrations measured at intersections. Mean concentrations were 56.1 ± 18.4 ppm for heavy traffic, 31.4 ± 31.5 ppm for moderate traffic, and 15.3 ± 10.2 ppm for light traffic. Ramsey (1966) also reported that concentrations were greater at intersections along major arteries somewhat removed from downtown Dayton, and that their mean concentration was 3.4 times the mean of intersections a block away and perpendicular to the axis of the arterial. Claggett *et al.* (1981) found that CO concentrations at intersections with signals were higher than those measured near freeways that had two to three times greater traffic volumes.

Others reported similar results for VOCs. Seifert and Abraham (1982), who surveyed VOCs near traffic intersections of German cities, reported mean concentrations in 100 μg m^{-3} for m-/p-xylene and 147 μg m^{-3} for toluene, and 12–193 μg m^{-3} for benzene.

4.2.4 Service stations

In some places, people serve themselves with fuel at service stations. This practice potentially puts these people at risk of exposure to evaporative emissions during refuelling and other motor vehicle pollutants.

Amemdale and Hanes (1984) measured CO concentrations in 13 automobile service stations and two dealer shops in the New England area of the USA. Results varied by season, with CO concentrations ranging from 2.2 to 21.6 ppm in warm weather and from 16.2 to 110.8 ppm during cold weather. They cited reduced ventilation as another cause of higher concentrations during winter. Wilson *et al.* (1991) sampled at random 100 self-service filling stations, and for comparison took convenience samples at 10 parking garages and 10 office buildings near the garages in the Los Angeles, Orange, Riverside and San Bernardino Counties of Southern California. They took 5-minute samples of 13 motor vehicle air pollutants in each microenvironment and in the ambient environment. Table 4.6 summarises these results. Service station concentrations ranked second to parking garage concentrations for most pollutants.

4.2.5 Confined spaces

Studies have shown very high concentrations of CO and other pollutants in poorly ventilated, confined spaces used by motor vehicles. These include parking garages, tunnels and underpasses.

Table 4.6 Median 5-minute concentrations of motor vehicle air pollutants in Southern California, 1990

Pollutant	Service station	Parking garage	Office building	Ambient air
Benzene (ppb)	9	21	5	4
Carbon monoxide (ppb)	4,300	11,000	4,000	2,000
Ethylene dichloride (ppb)	0.1	0.1	0.1	0.1
Formaldehyde (ppb)	4	34	36	20
m-/p-xylenes (ppb)	13	43	9	9
Toluene (ppb)	36	49	26	29

Source: Wilson *et al.*, 1991

Parking garages

Initial studies of parking garages focused primarily on the garage itself. Trompeo *et al.* (1964) surveyed 12 underground garages in Turin, Italy and reported that CO concentrations averaged 98 ppm and ranged from 10 to 300 ppm for 132 observations. Chovin (1967) measured CO concentrations of 80–100 ppm on average in poorly ventilated garages in Paris, France. Goldsmith (1970) found that large numbers of cars queuing to leave parking garages could elevate pollutant levels inside garages to extremely high concentrations. Yocum *et al.* (1971) monitored Pb particles and CO concentrations in buildings in Hartford, Connecticut. Although one building was built over a parking garage, its ventilation system, which operated at a slight positive pressure, was effective in preventing the penetration of Pb particles from automobile exhaust into the building.

More recent studies have focused on the diffusion of air pollutants from garages into adjacent but unprotected office buildings and shopping centres. Flachsbart and Ott (1986) observed diffusion of CO from an underground parking garage into the upper floors of a 15-story office building in Palo Alto, California. A survey of the entire building showed that average CO concentrations in the garage were 40.6 ppm, and ranged from 10.2 to 18.5 ppm on the first 11 floors. Average CO concentrations on the top four floors ranged from 2 ppm to 4 ppm, because they were served by a separate ventilation system. The researchers attributed the higher CO levels to two factors:

- Garage fans were shut down to reduce electricity costs because rates had increased; and
- Doors connecting the main stairwell and garage of the building were left open.

Once these factors were corrected, average CO concentrations in the garage dropped from 40.6 to 7.9 ppm, while CO concentrations in the building fell to 0.7–1.5 ppm.

These findings in California described above motivated Flachsbart and Brown (1989) to survey a sample of 25 business outlets at a shopping centre attached to a semi-enclosed parking structure in Honolulu, Hawaii. At ground level the structure had a two-lane driveway lined by shops, most of which left doors open during business hours. An estimated 29.2 per cent of employees were exposed to CO levels in excess of the National Ambient Air Quality Standard of 9 ppm for 8 hours. The average CO concentrations for 30 visits to the sampled shops over the 5-month survey ranged from 2.8 to 23.1 ppm. Readings of CO at the nearest fixed-site monitor were usually below 1 ppm, and satisfied both state and federal ambient air quality standards. An earlier study of the same shopping centre by Bach and Lennon (1972) found average CO concentrations ranging from 12 to 37 ppm in parking areas. Flachsbart and Brown (1989) attributed their findings of lower CO levels to the wider use of more effective emission controls on automobiles which were in effect when they performed their study.

Tunnels

Early studies of tunnels found high pollutant concentrations similar to those found in garages. Waller *et al.* (1961) found high concentrations in London's Blackwell Tunnel and Rotherhithe Tunnel. Particulate matter ranged from 93 to 235 $\mu g\ m^{-3}$ and NO_X ranged from 1 to 8 ppm. The average CO concentrations during the morning and evening rush hours were slightly above 100 ppm. Peak CO concentrations reached 340–500 ppm on days when fans were shut off. Chovin (1967) found that pollutant concentrations in a tunnel depended on the length of the tunnel. Ayres *et al.* (1973), who studied New York tunnels, reported mean CO concentrations of 63 ppm for a 30-day period. Lead concentrations averaged 30.9 $\mu g\ m^{-3}$ with peaks of 98 $\mu g\ m^{-3}$.

More recently, Coviaux *et al.* (1984) measured CO concentrations in a tunnel in Paris, France. They found average CO concentrations of 30 ppm in the tunnel itself and 6 ppm in a "technical room" that the supported operation of the tunnel (however, smokers were present in the room). Lonneman *et al.* (1986) reported that measurements of non-methane organic carbon taken in 1982 in New York's Lincoln Tunnel were one quarter of the levels found in a 1970 study of the same site. They attributed this reduction to the greater prevalence of catalytic converters in the motor vehicle population, because tunnel ventilation rates for each year had not changed.

Underpasses
Wright *et al.* (1975) reported that average CO concentrations in six poorly ventilated underpasses in Toronto, Canada, ranged from 17.5 to more than 100 ppm.

4.3 Estimates of people exposed

This section provides estimates of the number of people exposed to air pollutants inside vehicles and along roadsides in developed and developing countries on a daily basis. Appendix 4.1 describes the methodology underlying these estimates. This methodology is based on the following assumptions:

- Human exposure to motor vehicle air pollution largely occurs in urban regions of the world. In comparison with rural areas, cities have more roadways, parking garages and street canyons where people may be exposed to high pollutant concentrations. Hence, a country's level of urbanisation is one indicator of the number of people exposed to motor vehicle air pollution. Nevertheless, some urban regions have good ambient air quality, as indicated by fixed-site monitors. Such air quality is partly due to favourable meteorological and topographical conditions. However, many studies reviewed in this chapter have shown that fixed-site monitors underestimate high personal exposures to some motor vehicle emissions. Thus, favourable meteorological and topographical conditions may offer no immunity from high personal exposures inside vehicles and along roadsides.
- The number of people exposed to air pollutants from motor vehicles depends upon the extent to which a country is motorised. Zahavi (1976) defined motorisation at the country level as the number of cars per 1,000 people or the number of cars per household. Countries with higher levels of motorisation can potentially lead to a greater numbers of people exposed to motor vehicle air pollutants.

The extent of motorisation in a particular country can be offset by the extent and level of emission controls on its motor vehicles. Renner (1989) briefly summarised the history of emission controls in different countries. In the USA, emission controls on motor vehicles were first implemented nationwide for new cars in the year 1968, and emission limits were gradually tightened in later years. Japan's emission standards, which are comparable to USA limits, were implemented in 1975 and 1978. More recently, Australia, Canada and South Korea adopted emission standards equivalent to those in the USA. The European Economic Community (EEC) adopted separate standards for large, medium and small vehicles. Small cars, which capture about 60 per cent of the European market, have the most lenient emission standards. Brazil expects to match current USA standards by 1997. Many other

Table 4.7 Definition of economic groups within countries

Economic group	Per capita GNP (1988 US$)
Low income	100–490
Lower-middle income	570–2,160
Upper-middle income	2,290–5,420
High income	6,010–27,500

Source: World Bank, 1990

countries, such as Argentina, India and Mexico, have virtually no controls on motor vehicle emissions.

- Urbanisation and motorisation vary by country and countries can be grouped according to the extent of their economic development. Based on standard definitions used by the World Bank (1990), four economic groups are defined here (Table 4.7) based on per capita gross national product (GNP).

4.3.1 Human and automobile populations

Table 4.8 summarises the world's population and motor car numbers for the four economic groups defined in Table 4.7, which include 159 countries representing 99.8 per cent of the world's total population of 5,313 million people in 1990 (note that this world population total includes about 20.5 million people living in Taiwan in 1990, but excluded from Table A.4 in United Nations (1991) which was the basis for the estimate of global population).

In 1990, while high-income countries had only 22.7 per cent of the globe's population, they had 37 per cent of the world's urban population of 2,400 million people and an astounding 85.6 per cent of the world's automobile fleet of 405 million vehicles. Middle-income countries, including both lower-middle and upper-middle categories, had 20.4 per cent of global population, but 26.3 per cent of the world's urban population and 12.7 per cent of all automobiles. By contrast, the low-income countries had 56.9 per cent of global population, but only 36.7 per cent of the urban population and only 1.7 per cent of the world's automobile fleet. The most frequent means of travel in low-income countries is by foot and forms of mass transit, such as pedal and motorised rickshaws, small minibuses, shared taxis, motorcycles and mopeds, and converted vans, pickups and jeeps (Armstrong-Wright, 1986; White, 1990).

Bleviss (1990) reported that shares of the world's automobile fleet are shifting among different countries and regions. The USA share of the world's automobile population fell from 77 per cent in 1930 to 35 per cent in 1986. By the year 2000,

Table 4.8 Population distribution and numbers of cars in different economic groups in world regions, 1990

Economic group and region	No. of countries	Population estimate (10^3)	Proportion of world population (%)	Urban population (10^3)	Proportion of world urban population (%)	No. of cars (10^3)	Proportion of world car fleet (%)
Low income							
Africa	31	430,390	8.12	117,277	4.89	1,133	0.28
Asia and the Middle East	13	2,581,811	48.68	759,987	31.68	5,516	1.36
Latin America and the Caribbean	2	7,309	0.14	2,115	0.09	68	0.02
Low income total	**46**	**3,019,510**	**56.94**	**879,379**	**36.66**	**6,717**	**1.66**
Lower middle income							
Africa	15	144,558	2.73	61,908	2.58	2,451	0.61
Asia and the Middle East	13	328,537	6.19	155,434	6.48	8,212	2.03
Latin America and the Caribbean	1	378,423	7.14	265,168	11.05	22,882	5.65
Europe	20	3,245	0.06	1,143	0.05	n.a.	n.a.
Pacific Islands	7	5,445	0.10	1,068	0.04	57	0.01
Lower middle income total	**56**	**860,208**	**16.22**	**484,721**	**20.21**	**33,602**	**8.30**
Upper middle income							
Africa	5	66,028	1.24	37,668	1.57	4,460	1.10
Asia and the Middle East	2	44,295	0.84	30,953	1.29	975	0.24
Latin America and the Caribbean	5	56,974	1.07	49,520	2.06	6,579	1.62
Europe	7	54,509	1.03	28,758	1.20	5,905	1.46
Upper middle income total	**19**	**221,806**	**4.18**	**146,899**	**6.12**	**17,919**	**4.42**
High income							
Asia and the Middle East	12	465,019	8.77	327,346	13.65	45,863	11.32
Pacific Islands	20	20,265	0.38	17,268	0.71	8,600	2.12
Europe	4	440,211	8.30	335,745	14.00	140,137	34.60
North America and the Caribbean	2	276,253	5.21	207,570	8.65	152,201	37.58
High income total	**38**	**1,201,748**	**22.66**	**887,929**	**37.01**	**346,801**	**85.62**
World total	**159**	**5,303,272**	**100.00**	**2,398,928**	**100.00**	**405,039**	**100.00**

n.a. Not applicable

Sources: United Nations, 1991; Hoffman, 1991

Europe will have more automobiles on the road than the USA (OECD, 1983). Although developing countries currently have small shares of the world's automobile fleet, countries in Asia and Latin America are expected to double or triple their automobile populations by the end of this century.

4.3.2 Numbers of vehicular journeys

Table 4.9 summarises the estimates of daily vehicular trips in urban areas by economic group and by region in 1990, further broken down by travel mode. This table shows the gross disparity in both the absolute numbers and the percentages of vehicular trips for developed and developing countries, and particularly between opposite ends of the income spectrum. This disparity in urban vehicular trips is even more pronounced when trips by automobile are compared with trips by public and public-type transport.

The high-income countries, which have only 37 per cent of the world's urban population of 2.4 million people, account for a disproportionately high share of the world's daily vehicular trips in urban areas. Of the 3,400 million vehicular trips taken every day in cities, over half take place in high-income countries and almost 72 per cent of these are by automobile. The lower-middle and upper-middle income countries combined, which represent 26.3 per cent of the world's urban population, account for only 21.1 per cent of all vehicular trips and 14.1 per cent of the automobile trips. By contrast, the low-income countries, which represent 36.7 per cent of the world's urban population, account for only 27.2 per cent of the world's vehicular trips and just over 14 per cent of its automobile trips in urban areas.

4.3.3 Numbers of roadside exposures

Exposure along roadsides is a function of the shape of the street and how people use streets to travel and to do business. In urban areas, the greatest potential for exposure often occurs in business and commercial districts. This potential is reduced in many developed countries, which have shopping centres that are designed to separate pedestrians and motor vehicles. By contrast, some cities in developing countries have sidewalks in commercial districts that are narrow, or non-existent. Proudlove and Turner (1990) describe typical street conditions in many developing countries cities. Street vendors and hawkers in developing countries take up so much sidewalk space that pedestrians are often forced to walk in the street between the sidewalk and moving motor vehicles.

Table 4.10 summarises the estimates of roadside populations by economic group and region. The table shows that cities in low-income countries have an estimated 62–103 million people who spend a considerable amount of their

Table 4.9 Estimates of daily vehicular trips made by different modes of travel within different economic groups and wolrd regions, 1990

Economic group and region	No. of urban vehicular trips per day (10^3)	Proportion of total urban vehicular trips per day (%)	No. of trips per day by car (10^3)	No. of trips per day by public or public-type transport (10^3)	Proportion of world total vehicular trips by car (%)	Proportion of world total vehicular trips by public and public-type transport (%)
Low income						
Africa	123,260	3.61	23,632	99,629	1.90	4.60
Asia and the Middle East	802,017	23.50	152,726	649,291	12.27	29.95
Latin America and the Caribbean	2,240	0.07	44	1,796	0.04	0.08
Low income total	**927,517**	**27.18**	**176,802**	**750,716**	**14.21**	**34.63**
Lower middle income						
Africa	67,185	1.97	13,833	53,352	1.11	2.46
Asia and the Middle East	173,528	5.08	38,167	135,361	3.07	6.24
Latin America and the Caribbean	295,507	8.66	74,465	221,042	5.98	10.20
Europe	1,251	0.04	298	954	0.02	0.04
Pacific Islands	1,174	0.03	246	928	0.02	0.04
Lower middle income total	**538,645**	**15.78**	**127,009**	**411,637**	**10.20**	**18.98**
Upper middle income						
Africa	45,362	1.33	11,663	33,699	0.94	1.55
Asia and the Middle East	39,353	1.15	8,308	31,045	0.67	1.43
Latin America and the Caribbean	60,306	1.77	18,172	42,134	1.46	1.94
Europe	36,477	1.07	10,565	25,913	0.85	1.20
Upper middle income total	**181,498**	**5.32**	**48,708**	**132,791**	**3.92**	**6.12**
High income						
Asia and the Middle East	600,508	17.59	185,839	414,673	14.93	19.13
Pacific Islands	31,467	0.92	18,995	12,472	1.52	0.58
Europe	651,125	19.08	336,256	314,869	27.01	14.52
North America and the Caribbean	482,190	14.13	351,179	131,011	28.21	6.04
High income total	**1,765,290**	**51.72**	**892,265**	**873,025**	**71.67**	**40.27**
World total	**3,412,950**	**100.00**	**1,244,784**	**2,168,169**	**100.00**	**100.00**

Source: Derived from data in Table 4.8 using the methodology described in Appendix 4.1

Table 4.10 Estimates of roadside populations by economic group and world region, 1990

Economic group and region	Urban labour force (10^3)	Roadside population (10^3)		Proportion of total roadside population (%)	
		Low estimate	High estimate	Low estimate	High estimate
Low income					
Africa	43,332	6,500	10,833	5.4	5.2
Asia and the Middle East	369,173	55,376	92,293	45.8	44.3
Latin America and the Caribbean	874	131	219	0.1	0.1
Low income total	**413,379**	**62,007**	**103,345**	**51.3**	**49.6**
Lower middle income					
Africa	20,602	3,090	5,151	2.6	2.5
Asia and the Middle East	54,940	8,241	13,735	6.8	6.6
Latin America and the Caribbean	106,818	16,023	26,704	13.3	12.8
Europe	553	83	138	0.1	0.1
Pacific Islands	394	59	99	0.0	0.0
Lower middle income total	**183,307**	**27,496**	**45,827**	**22.8**	**22.0**
Upper middle income					
Africa	11,618	1,743	2,904	1.4	1.4
Asia and the Middle East	13,009	1,951	3,252	1.6	1.6
Latin America and the Caribbean	18,459	2,769	4,615	2.3	2.2
Europe	12,665	1,900	3,166	1.6	1.5
Upper middle income total	**55,751**	**8,363**	**13,937**	**6.9**	**6.7**
High income					
Asia and the Middle East	164,833	9,874	18,931	8.2	9.1
Pacific Islands	8,728	436	873	0.3	0.4
Europe	149,726	7,486	14,973	6.2	7.2
North America and the Caribbean	104,267	5,228	10,449	4.3	5.0
High income total	**427,554**	**23,024**	**45,226**	**19.0**	**21.7**
World total	**1,079,991**	**120,890**	**208,335**	**100.0**	**100.0**

Source: Derived from data in Table 4.8 using the methodology described in Appendix 4.1

working day in roadside situations. Their roadside population is roughly 1.7 times as large as that found in all middle-income countries and 2.3–2.7 times as large as that found in high-income countries. A comparison of Tables 4.8 and 4.10 shows that the percentages of the world's roadside population held by both low and high-income countries roughly correspond to their respective percentages of the world's total population. For lower-middle and upper-middle income countries, the same comparison shows that their respective percentages of the world's roadside population roughly correspond to their respective percentages of the world's urban population.

4.4 Conclusions

Several conclusions have emerged from studies of human exposure to the exhaust and evaporative emissions of motor vehicles.

1. Fixed-site ambient air quality monitors do not characterise adequately human exposure to certain air pollutants, especially CO exposure in certain micro-environments. For CO, monitors in two USA cities have overestimated exposures of people with low personal exposure, and underestimated exposures of people with high personal exposure. Exposure estimates based on fixed-site monitors appear to represent average or typical exposures rather than low- or high-level exposures. Special studies are needed to complement data from fixed-site monitors and to characterise more fully the exposures of subpopulations with high-level exposures. Furthermore, high CO exposures may exist for some people even though fixed-site monitors show that a city complies with ambient air quality standards.
2. Moderate-to-high personal exposures to air pollutants emitted by motor vehicles occur in certain urban micro-environments. Such exposures occur for congested roadways, street canyons, tunnels, underpasses, service stations, drive-up facilities, parking garages, and some buildings attached to garages. Exposures in some of these micro-environments may exceed ambient air quality standards. Because concentrations of some air pollutants (e.g. CO) decline with greater distance from a road, passengers of vehicles and bicyclists usually have higher exposures to such pollutants. Pedestrians and street merchants along roadsides have slightly less and the general urban population the lowest exposure.
3. In-vehicle exposures vary depending on whether the direct or indirect approach is used to survey travellers. Furthermore, in-vehicle exposures vary from city to city within a given country. Within a given city, exposures also vary from one season to another, from one mode of

travel to another, from one roadway to another, and from peak to non-peak travel periods. For a given roadway, exposure may vary from one lane to another, especially if the roadway has priority lanes for shared cars, shared vans and express buses. For a given vehicle, a passenger's exposure may be very high if their vehicle has a defective exhaust system or the surrounding vehicles have uncontrolled emissions. Also, passenger exposure varies depending on what ventilation setting (i.e. air flow or air conditioning level within the car) is used.

4. Curb-side CO concentrations in the commercial districts of cities vary depending upon distance from the roadway, degree of enclosure formed by street canyons, and the daily and seasonal changes that affect traffic patterns, ambient wind speeds and temperatures. In the absence of indoor sources, CO concentrations in enclosed settings are similar to outdoor concentrations, but tend to lag behind the peak concentrations observed outdoors. However, indoor concentrations tend not to reach either the extreme high or the extreme low values that are found outdoors. The exceptions are certain commercial buildings that are attached to well used parking garages. When ventilation systems in these buildings are not functioning or are not operating properly, motor vehicle emissions from garages may infiltrate buildings and expose occupants to concentrations higher than ambient levels.
5. A country's level of economic development plays a significant role in human exposure to motor vehicle air pollutants. Both low- and high-income countries have similar numbers of people who live in urban areas. However, countries with high incomes have a larger percentage of the world's automobiles. Consequently, these nations have greater mobility and larger proportions of their populations who use automobiles to make trips. Thus, the potential for human exposure to automotive air pollutants is high in developed countries because these risk factors exist. However, tight emission controls and the use of unleaded fuel appear to offset these factors. For example, despite the growth in the automobile population and vehicle miles of travel that has occurred in the USA during the last 25 years, human exposure to CO and Pb has diminished substantially.
6. Occupational and commuter exposures to air pollutants from motor vehicles in some large cities of developing countries are severe. In Metro Manila, Philippines, jeepney drivers are exposed to concentrations of CO, Pb, SO_2 and SPM that are above WHO's guidelines. In Mexico City, commuters are presently exposed to levels of CO that are

higher than exposures observed in the USA, at present and prior to the adoption of nationwide emission controls. Countries that still have low levels of motorisation and no emission controls on cars appear to have lower commuter exposures than in some developed countries.

Currently, a relatively low level of motorisation in China appears to offset a lack of emission controls on cars, such that the exposure of bicyclists is lower than in England. Nevertheless, levels of motorisation are growing rapidly, especially in some countries in Asia and Latin America; this may eventually create exposure problems. Motorcycle commuters in Taipei City, Taiwan, for example are exposed to benzene concentrations that greatly exceed the exposures of motorists in Los Angeles, California. In addition, large cities in developing countries have much greater percentages of their informal labour forces living and working in the street than in cities in developed countries. Thus, the number of people exposed to motor vehicle air pollutants in roadside settings may be higher in developing countries.

4.5 References

Akland, G., Hartwell, T., Johnson, T. and Whitmore, R. 1985 Measuring human exposure to carbon monoxide in Washington, D.C. and Denver, Colorado, during the winter of 1982–83. *Environmental Science and Technology*, **19**, 911–918.

Amemdale, A. and Hanes, N. 1984 Characterization of indoor carbon monoxide levels produced by the automobile. In: B. Berglund *et al.* [Eds] *Indoor Air, Volume 4, Chemical Characterization and Personal Exposure,* NTIS PB85-104214. Swedish Council for Building Research, Stockholm, Sweden, 97–102.

Amiro, A. 1969 CO presents public health problems. *Journal of Environmental Health*, **32,** 83–88.

Armstrong-Wright, A. 1986 *Urban transit systems: guidelines for examining options*. Technical Paper Number 52, World Bank, Washington, D.C.

Atkinson, R. 1988 Atmospheric transformations of automotive emissions. In: A. Watson *et al.* [Eds] *Air Pollution, the Automobile and Public Health.* National Academy Press, Washington, D.C., 99–132.

Ayres, S., Evans, R., Licht, D., Briesbach, J., Reimold, F., Ferrand, F. and Criscitiello, A. 1973 Health effects of exposure to high concentrations of automotive emissions. *Archives of Environmental Health*, **27**, 168–178.

Bach, W. and Lennon, K. 1972 Air pollution and health at Ala Moana Shopping Centre in Honolulu. *Hawaii Medical Journal*, **31**, 104–113.

Bellin, P. and Spengler, J. 1980 Indoor and outdoor carbon monoxide measurements at an airport. *Journal of the Air Pollution Control Association*, **30**, 392–394.

Besner, D. and Atkins, P. 1970 The dispersion of lead and carbon monoxide from a heavily travelled expressway. Paper No. 70–12 presented at the 63rd Annual Meeting of the Air Pollution Control Association, St. Louis, Missouri.

Bevan, M., Proctor, C., Baker-Rogers, J. and Warren, N. 1991 Exposure to carbon monoxide, respirable suspended particulates, and volatile organic compounds while commuting by bicycle. *Environmental Science and Technology*, **25**, 788–791.

Bleviss, D. 1990 The role of the automobile. *Energy Policy*, **18,** 137–148.

Brice, R. and Roesler, J. 1966 The exposure to CO of occupants of vehicles moving in heavy traffic. *Journal of the Air Pollution Control Association*, **16**, 597–600.

Chan, C., Özkaynak, H., Spengler, J. and Sheldon, L., 1991a Driver exposure to volatile organic compounds, CO, ozone, and NO_2 under different driving conditions. *Environmental Science and Technology*, **25**, 964–972.

Chan, C., Spengler, J., Özkaynak, H. and Lefkopoulou, M. 1991b Commuter exposures to VOCs in Boston, Massachusetts. *Journal of the Air and Waste Management Association*, **41**, 1594–1600.

Chan, C., Lin, S. and Her, G. 1993 Student's exposure to volatile organic compounds while commuting by motorcycle and bus in Taipei City. *Journal of the Air and Waste Management Association*, **43**, 1231–1238.

Chan, L., Hung, W. and Qin, Y. 1994 Vehicular emission exposure of bicycle commuters in the urban area of Guangzhou, South China (PRC). *Environment International*, **20**, 169–177.

Chaney, L. 1978 Carbon monoxide automobile emissions measured from the interior of a travelling automobile. *Science*, **199**, 1203–1204.

Chovin, P. 1967 CO, analysis of exhaust gas investigations in Paris. *Environmental Research*, **1**, 198–216.

Claggett, M., Shrock, J. and Noll, K., 1981 Carbon monoxide near an urban intersection. *Atmospheric Environment*, **15**, 1633–1642.

Clements, J. 1978 *School Bus Carbon Monoxide Intrusion*. DOT–HS–803–705, National Highway Traffic Safety Administration, U.S. Department of Transportation, Washington, D.C.

Colucci, J. and Begeman, C. 1969 Carbon monoxide in Detroit, New York, and Los Angeles air. *Environmental Science and Technology*, **3**, 41–47.

Colwill, D. and Hickman, A. 1980 Exposurc of drivers to carbon monoxide. *Journal of the Air Pollution Control Association*, **30**, 1316–1319.

Cortese, A. and Spengle, J. 1976 Ability of fixed monitoring stations to represent personal carbon monoxide exposure. *Journal of the Air Pollution Control Association*, **26**, 1144–1150.

Coviaux, F, Mouilleseaux, A., Festy, B., Thibaut, G., Geronimi, J. and Viellard, H. 1984 Air quality and biological controls of workers exposed in working premises contiguous to an urban road-tunnel. In: B. Berglund, *et al.*, [Eds] *Indoor Air, Volume 4, Chemical Characterization and Personal Exposure.* NTIS PB85-104214. Swedish Council for Building Research, Stockholm, Sweden, 129–134.

Dor, F., Le Moullec, Y. and Festy, B. 1995 Exposure of city residents to carbon monoxide and monocyclic aromatic hydrocarbons during commuting trips in the Paris metropolitan area. *Journal of the Air and Waste Management Association*, **45**, 103–110.

Duan, N. 1982 Models for human exposure to air pollution. *Environment International*, **8**, 305–309.

Dzubay, T., Stevens, R. and Richards, L. 1979 Composition of aerosols over Los Angeles freeways. *Atmospheric Environment*, **13**, 653–659.

Fernandez-Bremauntz, A. 1993 Commuters' exposure to carbon monoxide in the metropolitan area of Mexico City, Doctoral dissertation, Centre for Environmental Technology, Imperial College of Science, Technology and Medicine, University of London, London.

Flachsbart, P. 1993 Covariates and models of personal exposure to emissions from motor vehicles. In: R. Manuell *et al.* [Eds] *International Workshop on Human Health and Environmental Effects of Motor Vehicle Fuels and Their Exhaust Emissions*. International Programme on Chemical Safety, UNEP–ILO–WHO, Sydney, Australia, 193–215.

Flachsbart, P. 1989 Effectiveness of priority lanes in reducing travel time and carbon monoxide exposure. *Institute of Transportation Engineers Journal*, **59**, 41–45.

Flachsbart, P. 1995 Long-term trends in United States highway emissions, ambient concentrations, and in-vehicle exposure to carbon monoxide in traffic. *Journal of Exposure Analysis and Environmental Epidemiology*, **5**, 473–495.

Flachsbart, P. and Ah Yo, C. 1989 *Micro Environmental Models of Commuter Exposure to Carbon Monoxide from Motor Vehicle Exhaust.* Environmental Monitoring Systems Laboratory, U.S. Environmental Protection Agency, Research Triangle Park, North Carolina.

Flachsbart, P. and Brown, D. 1989 Employee exposure to motor vehicle exhaust at a Honolulu shopping centre. *Journal of Architectural and Planning Research*, **6**, 19–33.

Flachsbart, P., Mack, G., Howes, J. and Rodes, C. 1987 Carbon monoxide exposures of Washington commuters. *Journal of the Air Pollution Control Association*, **37**, 135–142.

Flachsbart, P. and Ott, W. 1984 *Field Surveys of Carbon Monoxide in Commercial Settings Using Personal Exposure Monitors*, EPA–600/4–84–019. U.S. Environmental Protection Agency, Washington, D.C.

Flachsbart P. and Ott, W. 1986 A rapid method for surveying CO concentrations in high-rise buildings. *Environment International*, **12,** 255–264.

General Electric Company 1972 *Indoor-Outdoor Carbon Monoxide Pollution Study*. NTIS–PB220428, National Technical Information Service, Springfield, Virginia.

Godin, G., Wright, G. and Shephard, R. 1972 Urban exposure to carbon monoxide. *Archives of Environmental Health*, **25**, 305–313.

Goldsmith, J. 1970 Contribution of motor vehicle exhaust, industry, and cigarette smoking to community CO exposures. *Annals of the New York Academy of Science*, **15**, 122–134.

Haagen-Smit, A. 1966 Carbon monoxide levels in city driving. *Archives of Environmental Health*, **12**, 548–551.

Hoffman, M. [Ed.] 1991 *The World Almanac and Book of Facts*, Pharos Books, New York.

Holland, D. 1983 Carbon monoxide levels in micro environment types of four U.S. cities. *Environment International*, **9**, 369–378.

Jabara, J., Keefe, T., Beaulieu, H. and Buchan, R. 1980 Carbon monoxide: Dosimetry in occupational exposures in Denver, Colorado. *Archives of Environmental Health*, **35**, 198–204.

Joumard, R. 1991 Pollution de l'air dans les transports. *Pollution Atmosphérique*, **129**, 20.

Kleiner, B. and Spengler, J. 1976 Carbon monoxide exposures of Boston bicyclists. *Journal of the Air Pollution Control Association*, **26**, 147–149.

Koushki, P., Al-Dhowalia, K. and Niaizi, S. 1992 Vehicle occupant exposure to carbon monoxide. *Journal of the Air and Waste Management Association*, **42**, 1603–1608.

Lioy, P. 1990 Assessing total human exposure to contaminants: A multidisciplinary approach. *Environmental Science and Technology*, **24**, 938–945.

Lonneman, W., Sella, R. and Meeks, S. 1986 Non-methane organic composition in the Lincoln Tunnel. *Environmental Science and Technology*, **20**, 790–796.

Luria, M., Weisinger, R. and Peleg, M. 1990 CO and NO_X levels at the centre of city roads in Jerusalem. *Atmospheric Environment*, **24B**, 93–99.

Lynn, D., Ott, W., Tabor, E. and Smith, R. 1967 Present and future commuter exposures to carbon monoxide. Paper presented at the 60th Annual Meeting of the Air Pollution Control Association, Cleveland, Ohio.

Myronuk, D. 1977 Augmented ingestion of carbon monoxide and sulfur oxides by occupants of vehicles while idling in drive-up facility lines. *Water, Air, and Soil Pollution*, **7**, 203–213.

Nagda, N. and Koontz, M. 1985 Micro environmental and total exposures to carbon monoxide for three population subgroups. *Journal of the Air Pollution Control Association*, **35**, 134–137.

OECD 1983 *Long Term Outlook for the World Automobile Industry*. Organisation for Economic Co-operation and Development, Paris.

OECD 1988 *Cities and Transport*. Organisation for Economic Co-operation and Development, Paris.

Office of the National Environment Board 1989 *Air and Noise Pollution in Thailand 1989*. Air Quality and Noise Division, Office of the National Environment Board, Bangkok, Thailand.

Ott, W. 1982 Concepts of human exposure to air pollution. *Environment International*, **7**, 179–196.

Ott, W. and Eliassen, R. 1973 A survey technique for determining the representativeness of urban air monitoring stations with respect to carbon monoxide. *Journal of the Air Pollution Control Association*, **23**, 685–690.

Ott, W. and Flachsbart, P. 1982 Measurement of carbon monoxide concentrations in indoor and outdoor locations using personal exposure monitors. *Environment International*, **8**, 295–304.

Ott, W., Switzer, P. and Willits, N. 1993 Trends of in-vehicle CO exposures on a California arterial highway over one decade. Paper No. 93-RP-116B.04 presented at the 86th Annual Meeting of the Air and Waste Management Association, Denver, Colorado.

Ott, W., Switzer, P. and Willits, N. 1994a Carbon monoxide exposures inside an automobile travelling on an urban arterial highway. *Journal of the Air and Waste Management Association*, **44**, 1010–1018.

Ott, W., Switzer, P. and Willits, N. 1994b *Carbon Monoxide Exposures Inside an Automobile Travelling on an Urban Arterial Highway*. SIMS Technical

Report No. 150, Department of Statistics, Stanford University, Stanford, California.

Petersen, G. and Sabersky, R. 1975 Measurement of pollutants inside an automobile. *Journal of the Air Pollution Control Association*, **25**, 1028–1032.

Petersen, W. and Allen, R. and 1982 Carbon monoxide exposures to Los Angeles area commuters. *Journal of the Air Pollution Control Association*, **32**, 826–833.

Proudlove, A. and Turner, A. 1990 Street management. In: H. Dimitriou [Ed.] *Transport Planning for Third World Cities*. Routledge, London, 348–378.

Ramsey, J. 1966 Concentrations of carbon monoxide at traffic intersections in Dayton, Ohio. *Archives of Environmental Health*, **13**, 44–46.

Rashidi, M. and Massoudi, M. 1980 A study of the relationship of street level carbon monoxide concentrations to traffic parameters. *Atmospheric Environment*, **14**, 27–32.

Renner, M. 1989 Rethinking transportation. In: L. Starke [Ed.] *State of the World 1989*. W.W. Norton and Company, New York, 97–102.

Seifert, B. and Abraham, H. 1982 Indoor air concentrations of benzene and some other aromatic hydrocarbons. *Ecotoxicological and Environmental Safety*, **6**, 190–192.

Sexton, K. and Ryan, P. 1988 Assessment of human exposure to air pollution: methods, measurements and models. In: A. Watson *et al.* [Eds] *Air Pollution, the Automobile and Public Health*. National Academy Press, Washington, D.C., 207–238.

Shikiya, D., Liu, C., Kahn M., Juarros, J. and Barcikowski, W. 1989 *In-Vehicle Air Toxics Characterization Study in the South Coast Air Basin*. South Coast Air Quality Management District, El Monte, California.

Sisovic, A and Fugas, M. 1987 Survey of carbon monoxide concentrations in selected urban micro environments, *Environmental Monitoring and Assessment*, **9**, 93–99.

Sterling, T. and Kobayashi, D. 1977 Exposure to pollutants in enclosed "living spaces". *Environmental Research*, **13**, 1–35.

Subida, R. and Torres, E. 1991 *Epidemiology of Chronic Respiratory Symptoms and Illnesses Among Jeepney Drivers, Air-Conditioned Bus Drivers and Commuters Exposed to Vehicular Emissions in Metro Manila, 1990–91*. Department of Environmental and Occupational Health, College of Public Health, University of the Philippines, Manila.

Tonkelar, W. 1983 Exposure of car passengers to CO, NO, NO_2, benzene, toluene and lead. In: *Proceedings of the World Congress on Air Quality*, Paris, 329–335.

Trompeo, G., Turletti, G. and Giarusso, O. 1964 Concentrations of carbon monoxide in underground garages. *Rassegna di Medicina Industriale*, **33**, 392–393.

UNEP/WHO (United Nations Environment Programme and World Health Organization) 1994 *Air Pollution in the World's Megacities,* **36**, 4–13, 25–37.

Wallace, L. 1979 Use of personal monitors to measure commuter exposure to carbon monoxide in vehicle passenger compartments. Paper No. 79–59.2 presented at the 72nd Annual Meeting of the Air Pollution Control Association, Cincinnati, Ohio.

Waller, R., Commins, B. and Lawther, P., 1961 Air pollution in road tunnels. *British Journal of Industrial Medicine*, **18**, 250–259.

Webster, F., Bly, P., Johnston, R., Paulley, N. and Dasgupta, N. 1985 *Changing Patterns of Urban Travel.* European Conference of Ministers of Transport, Paris.

Weisel, C., Lawryk, N. and Lioy, P. 1992 Exposure to emissions from gasoline within automobile cabins. *Journal of Exposure Analysis and Environmental Epidemiology*, **4**, 79–96.

White, P. 1990 Inadequacies of urban public transport systems. In: H. Dimitriou, [Ed.] *Transport Planning for Third World Cities.* Routledge, London, 86–116.

WHO, 1987 *Air Quality Guidelines for Europe.* WHO Regional Publications, European Series No. 23, World Health Organization, Regional Office for Europe, Copenhagen.

Wilson, A., Colome, S. and Tian, Y. 1991 *Air Toxics Micro Environment Exposure and Monitoring Study.* South Coast Air Quality Management District, El Monte, California.

Wilson, C. and Schweiss, J. 1978a *Carbon Monoxide Study–Boise, Idaho, November 25–December 22, 1977, Part 2.* EPA–910/9–78-O55b. U.S. Environmental Protection Agency, Region X, Seattle, Washington.

Wilson, C. and Schweiss, J. 1978b *Carbon Monoxide Study–Seattle, Washington, October November 2, 1977, Part 2.* EPA-910/9-78-0546. U.S. Environmental Protection Agency, Region X, Seattle, Washington.

World Bank, 1990 *World Development Report 1990.* Oxford University Press, Oxford.

United Nations 1991 *World Urbanization Prospects 1990.* Department of International Economic and Social Affairs, United Nations, New York.

Wright, G., Jewczyk, S., Onrot, J., Tomlinson, P. and Shephard, R. 1975 Carbon monoxide in the urban atmosphere: hazards to the pedestrian and the street-worker. *Archives of Environmental Health*, **30**, 123–129.

Wright, G. [Ed.] 1991 *The Universal Almanac 1992*. Andrews and McMeel, New York, 345–476.

Yocum, J., Clink, W. and Cote, W. 1971 Indoor/outdoor air quality relationships. *Journal of the Air and Waste Management Association*, **21**, 251–259.

Yu, L., Hildemann, L. and Ott, W. 1994 *A Mathematical Model for Predicting Trends in Carbon Monoxide Emissions and Exposures on Urban Arterial Highways*, Department of Civil Engineering, Stanford University, Stanford, California.

Zahavi, Y. 1976 Travel characteristics in cities of developing and developed countries. World Bank Staff Working Paper No. 230, World Bank, Washington, D.C.

Ziskind, R., Rogozen, M., Carlin, T. and Drago, R. 1981 Carbon monoxide intrusion into sustained-use vehicles. *Environment International*, **5**, 109–123.

Ziskind, R., Fite, K., and Mage, D. 1982 Pilot field study: Carbon monoxide exposure monitoring in the general population. *Environment International*, **8**, 283–293.

Appendix 4.1

METHODOLOGY FOR ESTIMATION OF NUMBERS OF PEOPLE EXPOSED TO MOTOR VEHICLE POLLUTION

Separate methods have been used to estimate the numbers of people exposed inside vehicles and along roadsides.

Exposure within vehicles

In-vehicle exposure to air pollutants from motor vehicles is a function of several factors:

- The average speed, volume and composition of traffic, which varies according to the route and time of travel;
- The mode of travel and the ventilation of the vehicle;
- How long the trip takes.

These factors are expected to vary greatly among countries in ways that have not been thoroughly and systematically studied and compared for the world's urban population. For simplicity, it was assumed that the actual number of people exposed to motor vehicle air pollutants inside vehicles is primarily a function of how frequently a person uses a motor vehicle to make trips, i.e. personal mobility. Studies by Zahavi (1976) suggested that personal mobility could be estimated as a function of a country's degree of motorization. To estimate personal mobility in urban areas, a country's per capita gross national product (GNP) was used as a surrogate for its degree of motorization. This was done for two reasons: first, GNP data are more prevalent than motorization data, and second, a country's per capita GNP and degree of motorization were found to be highly correlated ($r = 0.887$, $p < 0.001$) for a convenience sample of 23 countries.

This sample of 23 countries was than augmented with mobility data for selected metropolitan areas in those countries (Table 4a). It was assumed that the mobility of each metropolitan area was representative of the mobility for all cities in the same country. A regression analysis of these data enabled development of the following model:

$$Y_{i1} = 1.033295 + 0.000066\,(X_{i1}) \qquad \text{(i)}$$

where: Y_{i1} = Urban vehicular trips per day per person of country i in the 1980s

X_{i1} = GNP per capita of country i in 1988 US$

Table 4a Daily vehicular trip per person and per capita GNP for selected metropolitan areas world-wide

Economic group	Metropolitan area	Vehicular trips per day per person	National per capita GNP (1988 US$)
Low income	Bombay, India	0.82	340
	Jakarta, Indonesia	0.77	440
	Karachi, Pakistan	1.76	350
	Lagos, Nigeria	0.30	290
Lower-middle income	Abidjan, Ivory Coast	1.05	770
	Bangkok, Thailand	1.12	1,000
	Bogota, Colombia	1.14	1,180
	Cairo, Egypt	0.67	660
	Mexico City, Mexico	1.73	1,760
	Sao Paulo, Brazil	1.52	2,160
Upper-middle income	Buenos Aires, Argentina	1.37	2,520
	Caracas, Venezuela	1.25	3,250
	Pusan, South Korea	1.24	3,600
	Seoul, South Korea	1.82	3,600
High income	Hong Kong	1.47	9,220
	Munich, Germany	2.31	18,480
	Osaka-Kobe-Kyoto, Japan	2.30	21,020
	Paris, France	1.84	16,090
	Sample of British towns	1.96	12,810
	Sample of French towns	2.14	16,090
	Sample of Spanish towns	1.12	7,740
	Singapore	1.41	9,070
	USA average	2.68	19,840

Sources: Webster *et al.*, 1985; Charles River Associates, 1988; OECD, 1988; Dimitrou, 1990; World Bank, 1990

This model was statistically significant ($F = 39.313$, $p < 0.001$) for 1, 21 degrees of freedom and the predictive power ($r^2 = 0.65$) was acceptable. This model was than used to estimate personal mobility by vehicles in urban areas of countries throughout the world.

Studies by Zahavi (1976) also suggested that a crude modal split estimate (i.e. per cent of urban trips by automobile versus public or public-type transport) could also be made based on the extent of motorization in a particular country.

Preliminary data analysis suggested that per capita GNP was also a good predictor of modal split, but that it was not as powerful a predictor as the level of motorisation in a country. Consequently, the following two models were developed for this purpose based on data from a convenience sample of 40 cities world-wide (Table 4b):

$$Y_{i2} = 22.034 + 0.001884 (X_{i2}) \quad \text{(ii)}$$

$$Y_{i2} = 18.825 + 0.097837 (X_{i2}) \quad \text{(iii)}$$

where: Y_{i2} = Percentage of urban trips by automobile of country i for the early 1980s

X_{i1} = GNP per capita of country i in 1988 US$

X_{i2} = Cars per 1,000 population of country i for the 1980s

The percentage of total vehicular trips by public or public-type transport for each country (Y_{i3}) was computed by subtracting the percentage of vehicular trips by automobile (Y_{i2}) from 100 per cent. For these variables, data for each country were available for different years throughout the 1980s. Data were not readily available for the same year for these three variables.

Equation (ii) was statistically significant ($F = 37.2$, $p < 0.001$) for 1, 38 degrees of freedom and equation [iii] was statistically significant ($F = 68.3$, $p < 0.001$) for (1, 38) degrees of freedom.

The predictive power ($r^2 = 0.49$) of equation (ii) was inferior to that ($r^2 = 0.64$) of equation (iii). Thus, equation (ii) was used to estimate modal splits for 21 (mostly small) countries for which data on motorisation were not available and equation (iii) was substituted for 138 countries for which motorization data were available. The assumption was made that these models, although based on a sample of 40 world cities, could be used to estimate modal splits for the entire urban population of other countries.

Roadside exposure

For the purpose of estimating how many people are exposed along roadsides in developing countries, street vendors and hawkers were assumed to be members of the informal subsector of the country's economy. According to Friedmann and Sullivan (1975), this includes handicraft workers (seamstresses, basket and mat makers, rope makers, silversmiths), street traders and service workers (peddlers, food vendors), casual construction workers (carpenters, bricklayers, plumbers, electricians) and "underground" occupations (prostitutes, professional beggars, police spies, dope peddlers, pickpockets). Friedmann and Sullivan (1975) stated that the informal sector provides employment for between 25 and 40 per cent of the urban economy in developing countries.

The proportion of the informal sector that "works the streets" in developing countries, and is thus exposed to motor vehicle air pollution, could not be found in the published literature. For the purposes of this study it was

Table 4b Modal split, motorisation and per capita GNP for selected metropolitan areas

Economic group	Metropolitan area	Percentage of urban trips by car	Cars per 1,000 population 1983–89	Per capita GNP (1988 US$)
Low income	Bombay, India	8	3	340
	Karachi, Pakistan	3	5	350
	Jakarta, Indonesia	27	6	440
	Nairobi, Kenya	45	5	370
Lower-middle income	Abidjan, Ivory Coast	33	15	770
	Amman, Jordan	44	54	1,500
	Ankara, Turkey	23	25	1,280
	Bangkok, Thailand	25	15	1,000
	Bogota, Colombia	14	26	1,180
	Cairo, Egypt	15	15	660
	Kuala Lumpur, Malaysia	37	72	1,940
	Manila, Philippines	16	13	630
	Medellin, Colombia	6	26	1,180
	Mexico City, Mexico	19	62	1,760
	Rio de Janeiro, Brazil	24	91	2,160
	Sao Paulo, Brazi	32	91	2,160
	Tunis, Tunisia	24	40	1,230
Upper-middle income	Budapest, Hungary	15	171	2,460
	Seoul, South Korea	9	22	3,600
High income	Amsterdam, Netherlands	80	350	14,520
	Barcelona, Spain	53	270	7,740
	Brussels, Belgium	43	374	14,490
	Copenhagen, Denmark	56	292	18,450
	Hamburg, Germany	50	364	18,480
	Liverpool, England	58	322	12,810
	London, England	61	322	12,810
	Madrid, Spain	37	270	7,740
	Milan, Italy	33	421	13,330
	Munich, Germany	61	364	18,480
	Naples, Italy	58	421	13,330
	Osaka, Japan	31	248	21,020
	Oslo, Norway	57	380	19,990
	Paris, France	63	409	16,090
	Singapore	47	100	9,070
	Stockholm, Sweden	48	393	19,300
	Stuttgart, Germany	44	364	18,480
	Tokyo, Japan	32	248	21,020
	USA	93	567	19,840
	Vienna, Austria	56	355	15,470
	Wellington, New Zealand	56	412	10,000

Sources: Webster *et al.*, 1985; Armstrong-Wright, 1986; OECD, 1988; Charles River Associates, 1988; World Bank, 1990; Hoffman, 1991

assumed that 15 or 25 per cent of the urban labour force represented an upper limit of the roadside population for developing countries. For developed countries, it was assumed that 5 to 10 per cent of the urban labour force was assumed to work in roadside settings. The following equation was used to make these estimates:

$$Y_{i4} = k\,(X_{i3})\,(X_{i4})$$

where: Y_{i4} = The number of people exposed to motor vehicle air pollution in roadside settings of country i in 1990

k = The proportion of the urban labour force working in roadside settings (i.e. 15 to 25 per cent in developing countries and 5 to 10 per cent in developed countries)

X_{i3} = The urban population of country i in 1990

X_{i4} = The percentage of the total population of country i in the labour force in 1988–90.

In the absence of adequate data, the assumption was made that the percentage of a country's total population in the labour force was similar to the percentage of its urban population in the labour force. Data on the percentage of the total population in the labour force came from UNDP (1991) and Wright (1991).

References

Armstrong-Wright, A. 1986 *Urban transit systems: Guidelines from examining options*. World Bank Technical Paper No. 52, World Bank, Washington D.C.

Charles River Associated, Inc., 1988 *Characteristics of Urban Transportation Demand.* Urban Mass Transportation Administration, U.S. Department of Transportation, Washington D.C.

Dimitriou H. 1990 Transport and third world city development. H. Dimitriou [Ed.] *Transport Planning for Third World Cities.* Routledge, London, 1–49.

Friedman, J. and Sullivan, F. 1975 The absorption of labor in the urban economy: the case of developing countries. In: J. Friedman and W. Alonso [Eds] *Regional Policy Readings in Theory and Applications*. MIT Press, Cambridge, MA, 473–501.

Hoffman M. [Ed.] 1991 *The World Almanac and Book of Facts*. Pharos Books, New York.

OECD 1988 *Cities and Transport*. Organisation for Economic Co-operation and Development, Paris.

UNDP (United Nations Development Programme) 1991 *Human Development Report*. Oxford University Press, New York.

Webster F. *et al.* 1985 *Changing Patterns of Urban Travel*. European Conference of Ministers of Transport, Paris.

World Bank 1990 *World Development Report*. Oxford University Press, Oxford.
Wright J. [Ed.] 1991 *The Universal Almanac*. Andrews and McMeel, New York, 345–476.
Zahavi Y. 1976 Travel characteristics in cities of developing and developed countries. Staff Working Paper No. 230, World Bank, Washington D.C.

Chapter 5*

MOTOR VEHICLE EMISSION CONTROL MEASURES

Generally, the goal of a motor vehicle pollution control programme is to reduce emissions from motor vehicles in use to the degree (within reason) necessary to achieve healthy ambient air quality for mobile source related pollutants in all areas of a city or country as rapidly as possible. For reasons of impossibility or impracticality, if this goal is not achievable, emissions shall be reduced to the practical limits of effective technological, economic, and social feasibility. Achievement of these alternative general goals requires more specific goals, namely: setting emission standards which, if met, would achieve the desired reductions, implementing programmes which enforce compliance with these standards, and controlling, where needed, vehicle usage. These emission reduction goals should be achieved in a manner which is equitable with respect to the population groups affected and, where direct trade-offs between alternative approaches exist, in the least costly manner.

Two ways of controlling motor vehicle emissions are to control emissions per vehicle mile travelled (VMT), and to control the number of vehicle miles travelled. These methods can be used together.

5.1 Controlling emissions per mile or per kilometre driven

Emissions per VMT may be controlled to varying degrees by controlling vehicle performance, and/or fuel composition (e.g. Pb, sulphur, volatility, oxygen content).

5.1.1 Vehicle performance

Vehicle performance may be controlled by regulating vehicle hardware and by assuring that vehicles in use are properly maintained. Vehicle hardware may be controlled by requiring vehicles to be designed and built to meet emission standards when new and during their useful life (arbitrarily defined as 5 years; 50,000 miles prior to 1990 in the USA although it is recognised that a vehicle's total lifetime is much longer if properly used and maintained)

* *This chapter was prepared by M.P. Walsh*

and to be modified if they do not. The emission standards of concern represent substantial reductions in emissions from uncontrolled vehicle levels, and therefore it is necessary that manufacturers consciously design their vehicles for emission control. Vehicle design may be evaluated by testing prototype vehicles. If individually constructed prototype vehicles are not capable of meeting emission standards, it is generally accepted that mass produced vehicles of the same design probably would not either. Conversely, if manufacturers are able to build prototypes which demonstrate that their designs are capable of meeting standards, the probability that the production vehicles will also meet standards may be increased.

Even when a manufacturer has "certified" his intended design, production vehicles may not meet standards (even when new) because either they are not constructed in all material respects to the same design as the prototype (i.e. they are misbuilt), or the manufacturer has failed to translate the design effectively into mass production. The first situation is legally equivalent to introduction into commerce of a non-certified design and may be guarded against by inspection and civil penalties as mentioned above. The second situation may result from "prototype-to-production slippage" and/or "production variability", depending on the circumstances. In such situations, the vehicles may not conform to the standards, but production vehicle testing is the only way their nonconformity may be detected. Therefore, testing of production vehicles is necessary if assurance is to be gained that production vehicles actually meet standards.

Vehicles may meet standards when new, but may fail to meet standards during their useful life, even though they have been maintained, due to excessive deterioration while in-use. Such deterioration may occur because the manufacturer failed to translate effectively the design into mass production, or because the certified design was inadequate because of the inability of the accelerated certification testing to simulate accurately in-use standards. In such cases, it is desirable that the manufacturer is required to fix the vehicle at his own expense for two reasons:

- To reduce emissions from the vehicle.
- To deter the manufacturer from building such vehicles in the future.

Mechanisms for identifying excessively polluting vehicles while in use include in-use testing, defect reporting by manufacturers and by vehicle owners, state inspection programmes, and assembly-line testing. Mechanisms for requiring manufacturers to fix such vehicles are provided by the recall and warranty programmes.

Even though vehicles may have been designed and built to meet standards, they will not do so unless properly maintained. To the extent that motor vehicles can be designed to eliminate or minimise necessary maintenance, the magnitude of this vital task of assuring proper maintenance can be reduced. A means for achieving reduced maintenance requirements might be to force technology by gradually requiring manufacturers to reduce maintenance performed during certification testing and to reduce similarly the prerequisite maintenance for warranty purposes. However, the inability of the certification process to assess the need for time-dependent (as opposed to mileage-dependent) maintenance suggests that warranty requirements would be a principal enforcement tool. The fundamental requirements for achieving proper maintenance of vehicles in use are:

- Providing the incentive for car owners to request proper maintenance.
- The ability and incentive for the market place to provide proper maintenance.

Requiring maintenance through a mandatory inspection/maintenance (I/M) programme would be the most effective "incentive" for car owners. A prerequisite for the success of an I/M programme, either in terms of maximising emission reduction benefits or in terms of obtaining public acceptance, is assurance to the public that manufacturers are "doing their part", i.e. that the vehicles are designed and built to meet standards if properly maintained, and that manufacturers will bear the cost if they do not meet standards.

A significant aspect of an I/M programme is the inspection test used to identify vehicles in need of maintenance. A short inexpensive test which correlates with the full test is highly desirable. Such a test would not only make the I/M programme more cost effective in terms of direct emission reduction capability, but it would also make the inspections an extremely valuable surveillance tool for use in conjunction with assembly-line testing, recall and warranty programmes. In addition, it would make I/M more acceptable by activating the performance warranty.

The ability of the marketplace to provide proper maintenance is first dependent on knowing what constitutes proper maintenance. Manufacturers are required to specify such "reasonable and necessary" maintenance in owner's manuals. Review of such instructions by Environmental Protection Agencies (EPAs) in order to assure that, at least, state of the art technology is reflected in reduced maintenance requirements can be important. Manufacturers may also be required to design their vehicles and maintenance requirements to avoid "foreseeable" instances of improper or misadjustment. Where particular fuels are required, such as unleaded petrol for catalyst-equipped vehicles, that fuel must be made available. Furthermore, measures must also be taken (such as warning labels in the filler inlet area and on the dashboard, and fuel filler inlet restrictors designed to accommodate only the slightly narrower fuel nozzle that should be used for unleaded fuel) to ensure that vehicle owners and drivers are fully aware of the need for unleaded fuel in their vehicles. Some consideration should also be given to pricing policies for unleaded and leaded fuels so that vehicle owners and drivers do not have an economic incentive to use leaded fuel when a vehicle requires unleaded fuel. The service industry must be educated to the requirements of tuning for emission control rather than for traditional performance and thus should be given sufficient incentive, such as the mandatory requirement on their customers of passing an I/M test to provide such tuning. The after-sales service and spare-parts industry must be able to provide replacement parts which do not adversely affect emission performance. In addition, measures to prevent intentional tampering with emission control systems must be undertaken.

5.1.2 Fuel composition

As discussed above, fuel composition may be controlled, as with unleaded fuel, for the purpose of proper maintenance. In such situations, the control of fuel composition is only an indirect measure for controlling emissions. Fuel composition may also be controlled as direct means of controlling emissions, such as when the Pb content of leaded petrol is reduced to control Pb emissions. Moreover, it should be noted, that unleaded petrol also lowers exhaust hydrogen carbon (HC) emissions compared with the use of leaded fuel because of changes in the characteristics of combustion chamber deposits. Studies carried out in the USA (Auto/Oil, 1990, 1991) indicate that exhaust HC, CO and NO_X emissions decrease significantly with lower fuel sulphur in petrol vehicles equipped with catalytic converters.

Petrol

Throughout much of the industrialised world, unleaded fuel has been the norm for more than a decade. Japan has been the world leader in this regard, with more than 90 per cent of its petrol being unleaded for about two decades. It was estimated that in 1994 approximately two-thirds of all petrol sold in the world was unleaded. Where leaded fuel is used, the Pb content should be reduced to no more than 0.15 g l^{-1}.

In addition to the reduction or elimination of Pb, it is possible to reformulate petrol, in order to reduce regulated and unregulated emissions of concern. As part of a comprehensive policy to reduce vehicle emissions, fuel reformulation can offset any increased risks associated with the introduction of unleaded petrol and complement the elimination of health risks due to Pb with an overall reduction in toxic emissions and the O_3 forming potential.

The potential for reformulating petrol to reduce pollutant emissions attracted considerable attention in the USA when pressure to adapt alternative fuels increased during the mid- to late 1980s. One result was a major co-operative research programme between the oil and automotive industries during the early 1990s. This was followed by a similar effort in Europe. The result is that a great deal has been learned about the potential for modifying petrol in a manner which can significantly improve air quality. An additional advantage of fuel reformulation is that it can reduce emissions from all vehicles on the road in much the same way that reducing Pb in petrol can reduce Pb emissions from all vehicles.

The most significant potential emission reductions that can be achieved by reformulation are reducing volatility (to reduce evaporative emissions), reducing sulphur (to improve catalyst efficiency), and adding oxygenated blend stocks (with a corresponding reduction in the high-octane aromatic hydrocarbons which might otherwise be required). The potential benefits of improving various fuel parameters are summarised below.

Lowering volatility. Control of petrol volatility is another important strategy for reducing vehicle emissions arising from evaporating and refuelling especially in areas with warmer climates. Experience in the USA indicated that when lead in petrol was reduced during the 1970s and 1980s, highly volatile butane components were added, at least in part, to enhance octane levels, with the result that petrol volatility and evaporative emissions increased significantly. Vehicle evaporating emissions account for roughly 30–50 per cent of total vehicle HC emissions. Control of fuel volatility is relatively inexpensive and easily accomplished and can lower such emissions substantially.

Fuel volatility, as measured by Reid Vapour Pressure (RVP) has a marked effect on evaporative emissions from petrol vehicles both with and without evaporative emission controls. Tests on vehicles without evaporative emission controls have shown that increasing the fuel RVP from 62 kPa (9 pounds per square inch (psi)) to approximately 82 kPa (12 psi) roughly doubled evaporative emissions (Mc Arragher, 1988). The effect is even greater in vehicles with evaporative emission controls: going from 62 kPa (9 psi) to 82 kPa (12 psi) RVP fuel increased average diurnal emissions (expansion and emission of vapour, mainly from the fuel tank, due to daily temperature variation) by more than five times, and average hot-soak emissions (evaporation of fuel, mainly from the carburettor bowl and petrol tank, each time the vehicle stops with a hot engine) by 25–100 per cent (US EPA, 1987). The large increase in diurnal emissions from controlled vehicles is due to saturation of the charcoal canister, which then allows subsequent vapours to escape to the air.

Vehicle refuelling emissions are also strongly affected by fuel volatility. In a comparative test on the same vehicle (Braddock, 1988), fuel with 79 kPa

(11.5 psi) RVP produced 30 per cent greater refuelling emissions than petrol with 64 kPa (10 psi) RVP (1.45 g l^{-1} compared with 1.89 g l^{-1} dispensed). In response to such results, the US EPA has established nation-wide summer-time RVP limits for petrol.

An important advantage of petrol volatility controls is that they can have an effect on emissions from vehicles already in-use and from the petrol distribution system. Unlike new-vehicle emissions standards, it is not necessary to wait for all vehicles to be replaced before they take effect. The emission benefits and cost-effectiveness of lower fuel volatility are greatest where few of the vehicles in use are equipped with evaporative controls. Even where evaporative controls are in common use, as in the USA, the control of fuel volatility may still be beneficial to prevent in-use volatility levels from exceeding those for which the controls were designed.

In its analysis of the RVP regulation, the US EPA (1987) estimated that the long-term refining costs of meeting a 62 kPa (9 psi) RVP limit throughout the USA would be approximately US$ 0.0038 per litre, assuming crude oil at US$ 20 per barrel. These costs were largely offset by credits for improved fuel economy and reduced fuel loss through evaporation, so that the net cost to the consumer was estimated at only US$ 0.0012 per litre.

Oxygenates. Blending small percentages of oxygenated compounds such as ethanol, methanol, tertiary butyl alcohol (TBA) and methyl tertiary-butyl ether (MTBE) with petrol has the effect of reducing the volumetric energy content of the fuel, and at the same time improving the antiknock performance, thus making possible a potential reduction in Pb and harmful aromatic compounds. If there is no change in the settings of the fuel metering system, lowering the volumetric energy content results in a leaner air-fuel mixture, and helps to reduce exhaust CO and HC emissions.

Up to 2.7 per cent MTBE can be added to petrol without any increase in NO_X. There are two opposing effects taking place with the addition of oxygenates: an increase in the ratio of air to fuel which tends to raise NO_X, and a lowering of flame temperatures which tend to reduce NO_X. With MTBE levels above 2.7 per cent, the lower flame temperature effect seems to prevail.

Available data indicate that ethanol can be added to petrol at levels as high as 2.1 per cent oxygen without significantly increasing NO_X concentrations; however, in excess of 2.1 per cent, NO_X levels could increase significantly. For example, US EPA test data on over 100 cars indicated that oxygen levels of 2.7 per cent or more could increase NO_X emissions by 3–4 per cent (US EPA, Personal Communication). The Auto/Oil study concluded that there was a statistically significant increase in NO_x of about 5 per cent with the addition of 10 per cent ethanol (3.5 per cent O_2) (Auto/Oil, 1990).

Ethyl tertiary butyl ether (ETBE) is a potential source of oxygenates but, there is insufficient data available at present, to determine its impact on NO_x emissions. The Auto/Oil study found about a 6 per cent increase in NO_x but the results were not statistically significant (Auto/Oil, 1990).

Other fuel variables. Lowering sulphur in petrol lowers emissions of CO, HC and NO_x from catalyst equipped cars. A regression analysis performed by the Auto/Oil study showed that the sulphur effect (lowered emissions) was significant for HC on all 10 cars, for CO on 5 cars, and for NO_x on 8 cars. There were no instances of a statistically significant increase in emissions, (Auto/Oil, 1991). Because oxygenates are sulphur free, their addition would tend to lower petrol sulphur levels. The result suggested that NO_x would go down about 3 per cent per 100 ppm of sulphur reduction.

The Auto/Oil study, investigated other fuel variables and found that NO_x emissions were lowered by reducing olefins, raised when boiling temperature for 90 per cent of the fuel (T90) was reduced, and only marginally increased when aromatics were lowered (Auto/Oil, 1990). In general, reducing aromatics and T90 caused statistically significant reductions in exhaust mass NMHC and CO emissions. Reducing olefins increased exhaust mass NMHC emissions; although, the O_3 forming potential of the total vehicle emissions was reduced (Colucci and Wise, 1992).

The reduction of aromatics from 45 per cent to 20 per cent caused a 42 per cent reduction in benzene but a 23 per cent increase in formaldehyde, a 20 per cent increase in acetaldehyde and about a 10 per cent increase in 1,3-butadiene. Reducing olefins from 20 per cent to 5 per cent resulted in about a 31 per cent reduction in 1,3-butadiene but had little significant effect on other toxic compounds. Lowering the T90 from 360 to 280 °F resulted in statistically significant reductions in benzene, 1,3-butadiene (37 per cent), formaldehyde (27 per cent) and acetaldehyde (23 per cent) (Auto/Oil, 1990).

Diesel fuel

Diesel fuel generally contains much higher levels of sulphur than petrol. Therefore, many OECD countries have decided to take steps to lower the sulphur levels as a strategy to reduce diesel particulate emissions. This has the added benefit of increasing the potential for catalytic control of diesel particulate and organic HC emissions.

Attention is now being turned to the possibilities for modifying diesel fuel composition as a quick and cost effective means of reducing emissions from existing vehicles. The two modifications which show the most promise are a reduction in sulphur content, and a reduction in the fraction of aromatic hydrocarbons in the fuel.

Sulphur content. In addition to a direct reduction in emissions of SO_2 and sulphate particles, reducing the sulphur content of diesel fuel reduces the indirect formation of sulphate particles from SO_2 in the atmosphere. In Los Angeles, it has been estimated that each gram of SO_2 emitted results in about one gram of sulphates in the atmosphere. Therefore, the indirect sulphate emissions due to SO_2 from diesel vehicles are about as great as their direct SPM emissions. Conversion of SO_2 to sulphates is highly dependent on local meteorological conditions, thus resulting in varying effects in different cities.

Aromatic hydrocarbons. A reduction in the aromatic hydrocarbon content of diesel fuel may also help to reduce emissions, especially where fuel aromatic levels are high. For existing diesel engines, a reduction in aromatics from 35 per cent to 20 per cent by volume would be expected to reduce transient SPM emissions by 10–15 per cent and NO_x emissions by 5–10 per cent. Hydrocarbon emissions, and possibly the mutagenic activity of the SPM soluble organic fraction, would also be reduced. Modelling studies of the refining industry have shown that reductions of this magnitude in aromatics can often be obtained by alterations in diesel fuel production and blending strategy, without a need for major new investments in additional processing capacity. Reduced diesel fuel aromatic content would also have other environmental and economic benefits. The reduced aromatic content would improve the fuel's ignition quality, leading to improved cold starting and idling performance and reduced engine noise. The reduction in the use of catalytically cracked blending stocks should also have a beneficial effect on deposit forming tendencies in the fuel injectors, thereby reducing maintenance costs. On the negative side, however, the reduced aromatics might result in some impairment of cold flow properties, due to the increased paraffin content of the fuel.

Fuel additives. A number of well controlled studies have demonstrated the ability of detergent additives in diesel fuel to prevent and remove injector tip deposits, thus reducing smoke levels. The reduced smoke probably also results in reduced SPM emissions, but this has not been demonstrated clearly, because of the great expense of SPM emissions tests on in-use vehicles. Cetane improving additives will also possibly result in some reduction in HC and SPM emissions in marginal fuels.

Table 5.1 Component control costs and VOC emission reductions

Component	Control level	Incremental cost (cents/litre)	Cumulative VOC reduction (%)
Oxygen	2.0 (Wt.%)	0.44–0.89 (based on MTBE)	9.0
Benzene	1.0 (Vol.%)	0.18	9.0
RVP	8.1 (psi)	0.15	17.6
RVP	7.4 (psi)	0.44	25.3
Sulphur	160 (ppm)	0.09–0.15	26.4
Oxygen	2.7 (Wt.%)	0.16–0.31	28.5
Olefins	5.0 (Vol.%)	0.48–0.64	30.2
Sulphur	50 (ppm)	0.38–0.49	31.2
Aromatics	20 (Vol.%)	0.16–0.26 (based on MTBE)	31.4

MTBE Methyl tertiary-butyl ether

Source: US EPA, 1994

Alternative fuels for buses

The possibility of substituting cleaner burning, alternative fuels for diesel fuel, has attracted increasing attention over the last decade. Purported benefits of this substitution include conservation of oil products and energy security, as well as the reduction or elimination of SPM emissions and visible smoke. The main alternative fuels presently under consideration are natural gas methanol made from natural gas and, in limited applications, liquid petroleum gasolines (LPG). Whether to use alternative fuels, and if so which fuel, requires a detailed study of the costs of implementing and sustaining the fuel supply system in a given location.

Cost effectiveness

It is difficult to estimate the costs and the cost effectiveness of fuel modifications because the characteristics of fuels produced by different refineries varies. Individual fuel component control costs, and the effects of changes in one fuel component on the other fuel components, are integral aspects of the determination of cost effectiveness. In an analysis by the US EPA, these two integral parts were estimated from the results of refinery modelling and survey results presented by the CARB.

The total cost (or manufacturing cost) of producing a reformulated gasoline (RFG) is the sum of the capital recovery cost and the operating cost. An example of the individual fuel component costs and the associated incremental per cent and reduction in VOC emissions are shown in Table 5.1.

A range of VOC and NO_X standards were proposed by the US EPA based on particular combinations of fuel component controls which reduce VOC (and VOC plus NO_X) emissions at a cost of less than US$ 5,000 per ton

(US$ 5,511 per tonne) and less than US$ 10,000 per ton (US$ 11,022 per tonne). These ranges were chosen to represent the upper limit of costs which would be incurred by many O_3 non-attainment areas in meeting standards.

Estimates of the costs and cost effectiveness of California reformulated petrol continue to fall. At the time it developed its regulations, CARB estimated the costs to be US$ 0.12–0.17 per gallon (US$ 0.03–0.04 per litre). Recently, a US EPA analysis placed the costs at US$ 0.08–0.11 per gallon (US$ 0.02–0.03 per litre). This analysis estimated the cost effectiveness of the California RFG to be US$ 4,100–5,100 per ton (US$ 4,519–5,622 per tonne) of VOC and NO_X control; Federal Phase 1 RFG was estimated to cost US$ 3,100 per ton (US$ 3,417 per tonne) of VOC control.

5.2 Controlling vehicle miles travelled

Reduction of VMT, e.g. through car sharing, increased use of mass transport, parking restrictions, petrol rationing, etc., is an additional means of controlling emissions. This can be achieved by the introduction of various transport policies such as:

- *Policies to induce shifts to more efficient modes of transportation.* These policies reduce transportation energy consumption and emissions per seat mile. Incentives include transit fare reductions and service improvements, such as extending coverage, reducing headways, improving travel time and the reliability of on-time service, co-ordinating transfers, and construction of park and ride facilities. Disincentives usually fall on the private automobile in the form of increased parking charges, surcharge taxes on motor fuel, and petrol and diesel fuel rationing.
- *Policies to increase the load factor of existing vehicle fleets.* These policies attempt to increase passenger miles per seat mile. Incentives include matching and information programmes for shared cars, shared van and commuter bus programmes, dial-a-ride, shuttle and jitney services. Examples of disincentives are similar to those noted above.
- *Policies to shift the time of peak travel demand occurrence.* These policies decrease transportation emissions per seat mile for all modes by spreading congested peak traffic loads over a broader time frame to improve the patterns of use for existing transportation capacity, thereby improving vehicle operating efficiency. Included among these policies are freeway ramp metering programmes, four day work weeks, staggered work hours and general traffic circulation improvements, such as synchronised signals.
- *Policies to reduce travel demand.* These policies attempt to decrease passenger miles through redistribution of urban activities. Examples include land-use policies to promote mixed land use, to increase density

along transit corridors, and to co-ordinate new subdivision development into efficient patterns. Technological examples include policies to encourage substitution of communications for transportation flows. Although such approaches are generally politically less acceptable, they tend to promote fuel conservation, and aid urban renewal. They may also represent the only mobile source control measure remaining once vehicle control technology has been pushed to its limits.

The success achieved in reducing per mile emissions from vehicles can eventually be eroded by continued high growth rates in the number and use of vehicles. With very few exceptions (Singapore and Curitiba among them), transportation controls to reduce such growth have failed, not because they cannot work but rather because most countries have not tried seriously to implement them.

It is now clear that technological solutions to the motor vehicle pollution problem are increasingly offset by growth in vehicle numbers. Therefore, the long-term solution of the environment problem depends on reducing the overall growth of transport. High growth effects emissions in two ways: it directly increases emissions (more miles driven leads to more pollution) and it leads to more congestion which further increases emissions.

5.3 A programme for reducing motor vehicle air pollution

The major elements of an overall vehicle pollution control strategy are reducing emissions per kilometre driven together with the amount of driving. Emissions per kilometre driven can also be lowered by altering some aspects of the driving itself such as average speed and degree of acceleration. A natural and consistent contradiction exists between altering driving characteristics and reducing the amount of driving because, frequently, strategies designed to increase average speed by improving traffic flow actually enable a given roadway network to carry more vehicles per hour. This effectively increases overall vehicle emissions. The use of these approaches together can be used to help ameliorate, otherwise likely, future pollution increases. As an example, an assessment of two representative cities is presented below.

5.3.1 An inventory of emissions

The potential for reducing emissions and improving air quality must begin with an assessment of existing vehicle emissions. Many factors affect the total inventory of motor vehicle emissions. Understanding these factors helps a better total inventory to be structured of these emissions and to determine optimal programmes for their control. Having inventories that reflect different control measures accurately, allows evaluation of the effectiveness of a given regulatory programme. Some of the more important factors to be included in a total inventory are:

- *Emission factors for new vehicles.* Ideally these emission factors can be determined from the emission standards for new vehicles subject to regulation, or from emission data available for similar vehicles (e.g. similar vehicle design, control technologies) used in other countries.
- *Deterioration of vehicle emissions with vehicle age and mileage.* Estimating how vehicle emissions deteriorate with time and mileage is critical in assessing in-use emissions. Different types of vehicles with different technologies (such as non-catalyst petrol fuelled, catalyst equipped, diesel fuelled), deteriorate differently with increased mileage and time. This type of information has been determined from the testing of in-use vehicles carried out by several different countries.
- *Tampering effects.* This adjustment accounts for vehicle owners or drivers intentionally altering or disabling an emission control system. Examples are disconnecting air pumps, catalysts, evaporative emission control systems, exhaust gas re-circulating systems and ignition timing. Vehicles are sometimes tampered with in the mistaken belief that vehicle performance, fuel economy or other factors (e.g. maintenance time or costs) will be improved. Information on the effects of tampering with vehicle emissions can be obtained from various countries. However, the incidence of vehicle tampering should be estimated for the particular country in which the inventory is being developed.
- *Vehicle maintenance.* If a vehicle (with or without emission controls) is not maintained according to the manufacturer's recommendations (e.g. periodic engine tuning, replacement of spark plugs or emission control components, carburettor adjustments) it will have significantly higher emissions than one which has been properly maintained. These higher emissions must be accounted for. Some initial estimates can be made based on similar information available from in-use vehicle emission levels measured in other countries.
- *Inspection and maintenance and anti-tampering checks.* The presence of an effective inspection and maintenance programme that identifies high emitting vehicles and ensures that they are repaired (or emission control systems replaced/repaired if tampered with) can help eliminate excess emissions resulting from the previous two factors. Estimates of the benefits of such programmes are available from tests conducted in different countries.
- *Technology mix.* The proportion of vehicles using different technologies (e.g. diesel, non catalyst, oxidation catalyst, 3-way catalyst, etc.) is critical in estimating total vehicle emissions. The proportion of technologies used

in a given country (and their appropriate emission factors) must be known or estimated.

- *Vehicle age*. The number of vehicles of a given age is important because older and higher mileage vehicles usually have higher emissions. This type of data can generally be available from vehicle registrations obtained by most governments.
- *Number of vehicles*. The total number of vehicles of a given model year in a given area is generally obtained through vehicle registration data and it must be known to calculate an inventory. A critical element is to make accurate projections for future years.
- *Vehicle miles travelled per vehicle per year*. The number of miles a given type of vehicle travels per year must be known. This number usually varies with vehicle age; older vehicles travel fewer miles a year. The number of miles travelled will be different from one country to another.
- *Vehicle misfuelling*. If both leaded and unleaded petrol are sold in an area where some vehicles require the use of unleaded fuel to protect the catalytic converter, the proportion of catalyst vehicles which incorrectly use leaded petrol must be determined.
- *Fuel characteristics*. Fuel volatility can be important determinant in determining vehicle evaporative emissions (which can account for as much as half of total hydrocarbon emissions). Other fuel characteristics such as sulphur content, distillation characteristics and oxygen content may also be important.
- *Ambient temperature*. The average daily temperature (generally maximum and minimum) must be known to predict vehicle emissions. Generally, separate inventories are calculated for warm weather conditions (when O_3 concentrations are at their peak) and cold weather (when CO concentrations are high).

5.3.2 Enforcement tools for emission controls

The certification process which requires testing of prototype cars prior to production can affect vehicle design at low mileage and, to a limited degree, the durability of emission controls. It can also affect vehicle maintenance and reduce the potential effectiveness of recall and warranty — review of manufacturer-proposed maintenance schedules can restrain the manufacturer from requiring excessive maintenance. It can also constrain the manufacturer to require less maintenance than was performed on a certification prototype. Some prototype maintenance, however, is not required to be recommended to the consumer. Its major advantage, that it affects vehicle design early in the design process and before actual production begins, is also its major weakness, i.e. that it deals inherently with somewhat artificial, prototype cars in an artificial environment. By its very nature, therefore, it cannot address production problems deterioration due to age, or real-world driving and ambient extremes, or the amount and quality of maintenance that will actually be performed in-use.

Assembly line testing test new production vehicles is the only technique which can be used to assure that vehicles when built, are in fact meeting emission standards. However, before they are sold the impact of assembly line tests on durability of design depends on requiring allowances for deterioration which are derived from other programmes, such as certification or in-use testing. Further, like certification, it cannot affect the amount or quality of maintenance performed in-use.

Recall and warranty programmes can provide some incentive to individuals to maintain their vehicles properly and are the only programmes which can directly affect the actual in-use durability of vehicles. These programmes are subject to the limitations of dealing only with properly maintained vehicles and with a generally less than perfect response on the part of individual vehicle owners. Such programmes appear only to effect vehicles already in use and which have been polluting to an excessive degree. They are, therefore, remedies after the event. Furthermore, much of the potential effectiveness of these programmes is lost once the vehicles have been in-use for more than one or two years. However, the major effect of these manufacturer-directed in-use programmes is that they act as a significant deterrent to the design and/or manufacture of vehicles which will fail to comply initially or as a result of deterioration with actual use.

Inspection and maintenance programmes have been demonstrated to lower emissions from existing vehicles in two ways:

- By lowering emissions from vehicles which fail the test and are required to be repaired.
- By encouraging owners of vehicles to take proper care of them and to avoid the potential costs of penalties and of the repairing of vehicles which have been tampered with or fuelled incorrectly.

Based on all available data, it has been estimated that a well run inspection and maintenance programme is capable of very significant emission reductions, such as 25 per cent for HC and CO and about 10 per cent for NO_X. The less significant NO_X reductions reflect solely the lower tampering rates resulting from inspection and maintenance and anti-tampering programmes because at present there has been no focused effort to design inspection and maintenance programs specifically to identify and correct NO_X problems. It is also important to note that reductions in emissions begin slowly and gradually increase over time because I/M programmes tend to lower the overall rate of fleet emission deterioration. Maximum inspection and maintenance benefits are thereby achieved by adopting the programme as early as possible.

Inspection and maintenance is the only compliance technique which ensures that in-use vehicles are properly maintained. By requiring that vehicles pass a re-test, it has a direct impact on the quantity and quality of maintenance and also on the design through the warranty and recall programmes which use I/M as a surveillance tool.

Inspection/maintenance is probably the most effective "anti-tampering" programme because of the intensive surveillance built into the periodic inspection. Such surveillance is particularly helpful in addressing vehicle maladjustments which cause vehicles to exceed standards. However, where I/M is not in effect, gross tampering by dealers may be substantially deferred by an aggressive anti-tampering programme. The benefits of I/M, however, are limited by the adequacy of the short test used by the ability of the service industry to make proper repairs and by the potential tampering which could occur following the test and which could allow the vehicle to emit high emissions throughout the year. It seems clear, therefore, that the ideal programme must include all of the above elements.

5.3.3 Schedule and sequence for introducing controls

In a perfect world, with unlimited resources, it would be desirable to implement all of the above strategies, simultaneously, at the very start of a programme. This is very rarely, if ever, feasible or practical, and could even be counterproductive. Therefore, it is necessary to make decisions about how to phase-in various strategies. One such approach is to devote the early stages of the programme to laying a firm foundation for further action. Primary steps are:

- The adoption of necessary and feasible standards.
- Widespread distribution of necessary fuels.
- The imposition of limitations on importing vehicles, which have not been certified, solely to individuals (rather than to commercial companies).
- The imposition of limited certification and assembly line test programmes to ensure adequate vehicle designs.
- The introduction of voluntary I/M programmes (which mean every vehicle must be inspected but repair and re-test is not mandatory).
- The design and construction of necessary government testing facilities for future programmes.

In the second phase, the requirements can be gradually tightened. Inspection/Maintenance should become mandatory and the standards made progressively more stringent, importation of non-certified cars should be restricted, and selective government testing should be used to verify certification results and to institute necessary recalls.

Finally, as the fundamental structure becomes accepted, it can be gradually and routinely tightened. For manufacturers, standards can be made more stringent as technology advances. Recall liability and warranty protection can be extended to the actual vehicle lifetime, and allowable maintenance requirements can be reduced. Fuels can also continue to be improved. On board diagnostics can be introduced on new cars and I/M programmes can be modified to use them more fully.

5.3.4 Examples of applying different strategies

A typical breakdown of emissions for major urban areas in the Asia Pacific region, for example, would usually indicate that motorcycles are a major contributor to HC and organic SPM and a significant source of CO and Pb (see for example Figure 5.1 for Surabaya, Indonesia and Figure 6.1 for Bangkok). Diesel vehicles are the major source of sulphate and a significant source of carbonaceous SPM. Passenger cars dominate the CO and Pb emissions and contribute significantly to NO_X and hydrocarbons. Therefore, different vehicle categories must be the focus of attention to address different aspects of the overall vehicle pollution problem.

The resulting impact of various control strategies has been predicted for Surabaya and Bangkok (see Figures 5.2 and 6.2 respectively). An immediate priority should be to restrain future vehicle growth rates. Economic measures, physical restrictions and selective policies will, nevertheless, each play a role. However, even if overall growth in vehicle numbers could be constrained to only 5 per cent per year, well beyond the current average in many countries in the region, vehicle emissions would increase dramatically over the next 15 years. In addition to growth restraint, therefore, a series of additional strategies is necessary.

5.4 Advances in vehicle pollution controls

5.4.1 Petrol fuelled vehicles

Significant progress has occurred during the past two decades in the development of a wide variety of emission reduction technologies for petrol fuelled vehicles. Before controls were required, engine crankcases were vented directly to the atmosphere. Crankcase emission controls, which basically consist of closing the crankcase vent port, were introduced on new cars in the USA in the early 1960s with the result that, today, control of these emissions

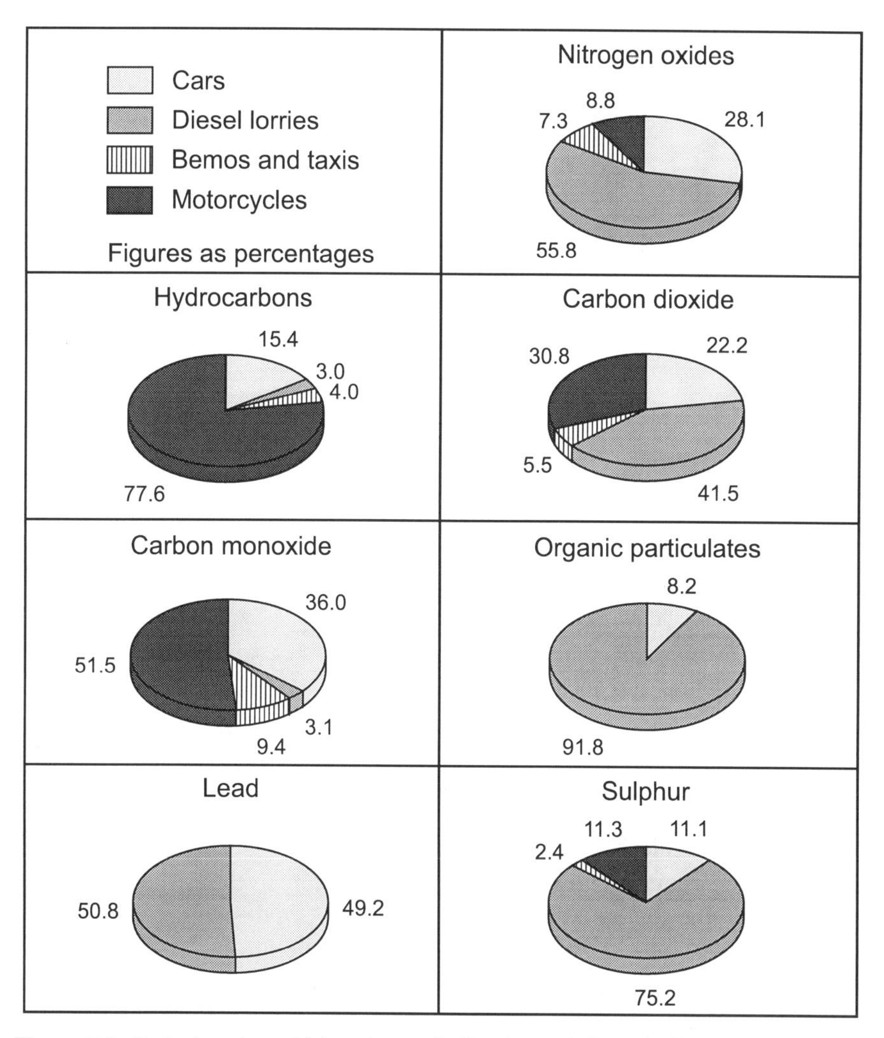

Figure 5.1 Emissions by vehicle category in Surabaya, Indonesia (Data provided by BAPEDAL, Indonesian Environment Agency, 1991)

is no longer considered a serious technical concern. Hydrocarbon evaporative emissions result from the distillation of fuel in the carburettor float bowl and the evaporation of fuel in the petrol tank. To control these emissions, manufacturers generally feed them back into the engine to be burned together with the other fuel. When the engine is not in operation, vapours are stored, either in the engine crankcase or in charcoal canisters which have a strong affinity for these emissions. They are then burned when the engine is started (such an approach improves vehicle fuel efficiency because vapours that

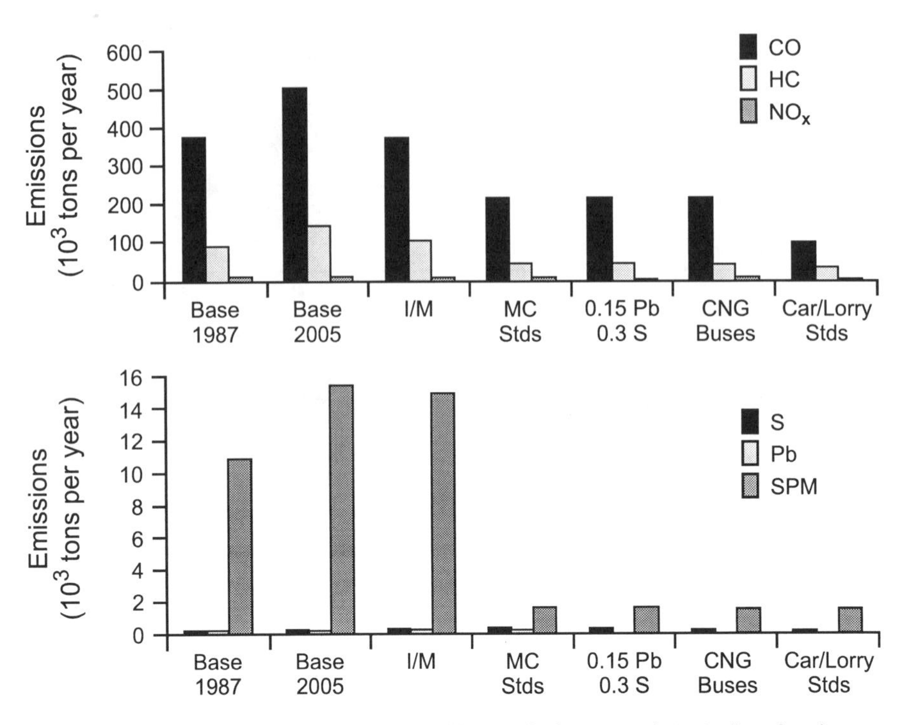

Figure 5.2 Potential impact of different vehicle emission control strategies, Surabaya, Indonesia (Data provided by BAPEDAL, Indonesian Environment Agency, 1991)

were previously released to the atmosphere are put to work). Technology to control these vapours is now readily available but its overall effectiveness is dependent on the fuel volatility for which it has been designed.

Exhaust emissions of HC, CO and NO_X are related to the air/fuel mixture inducted, the peak temperatures and pressures in each cylinder, whether lead is added to the petrol, combustion chamber geometry, and other engine design parameters. Variations in these parameters are, therefore, capable of causing significant increases or decreases in these emissions. The most important parameters are probably air/fuel ratio and mixture preparation, ignition timing, and combustion chamber design. Variations in these parameters are usually key elements in vehicle controls when modest light duty vehicle emissions standards are imposed. Dilution of the incoming charge has been shown to reduce peak cycle temperature by slowing flame speed and absorbing some heat of combustion. Re-circulating a portion of the exhaust gas back into the incoming air/fuel mixture (exhaust gas re-circulation, EGR) thereby lowering peak cycle temperature, is therefore used to lower NO_X emissions.

Improvements in mixture preparation, induction systems, and ignition systems can increase dilution tolerance. Tolerance can also be increased by increasing the burn rate or flame speed of the air-fuel charge. Techniques to do this include increased swirl and squash, shorter flame paths and multiple ignition sources.

Electronics

With so many interrelated engine design and operating variables playing an increasingly important role in the modern engine, the electronic control system has taken on increased importance. Modifications in spark timing must be closely co-ordinated with air-fuel ratio changes and degrees of EGR just in case significant fuel economy or performance penalties result from emissions reductions, or NO_X emissions increase as CO decrease. In addition, controls which can be much more selective, depending on engine load or speed, have been found beneficial in preventing widespread adverse impacts. Therefore, electronic controls have begun to replace more traditional mechanical controls. For example, electronic control of ignition timing has demonstrated an ability to optimise timing under all engine conditions and has the added advantage of reduced maintenance and improved durability, when compared with mechanical systems. When coupled with electronic control of EGR, it has been demonstrated that NO_X emissions can be reduced without penalties in fuel economy and in some cases with improved fuel economy.

Exhaust after-treatment devices

When stringent exhaust emission standards (especially HC or NO_X) are mandated, exhaust after-treatment devices such as catalytic converters tend to be used to supplement engine modifications. An oxidation catalyst is a device which is placed on the exhaust pipe of a car and which, if the chemistry and thermodynamics are correct, will oxidise almost all the HC and CO in the exhaust stream.

Three-way catalysts (so called because of their ability to lower HC, CO and NO_X levels simultaneously) were first introduced in the USA in 1977 by the Volvo company and subsequently became widely used when the NO_X standard in the USA was lowered to 1.0 gram per mile. To work effectively, these catalysts require precise control of air-fuel mixtures. As a result, three-way systems have indirectly fostered improved air-fuel management systems, such as advanced carburettors and throttle body fuel injection systems, as well as electronic controls.

Starting with cars manufactured in 1975 catalysts have been placed in more than 80 per cent of all new cars sold in the USA, and since 1990 either

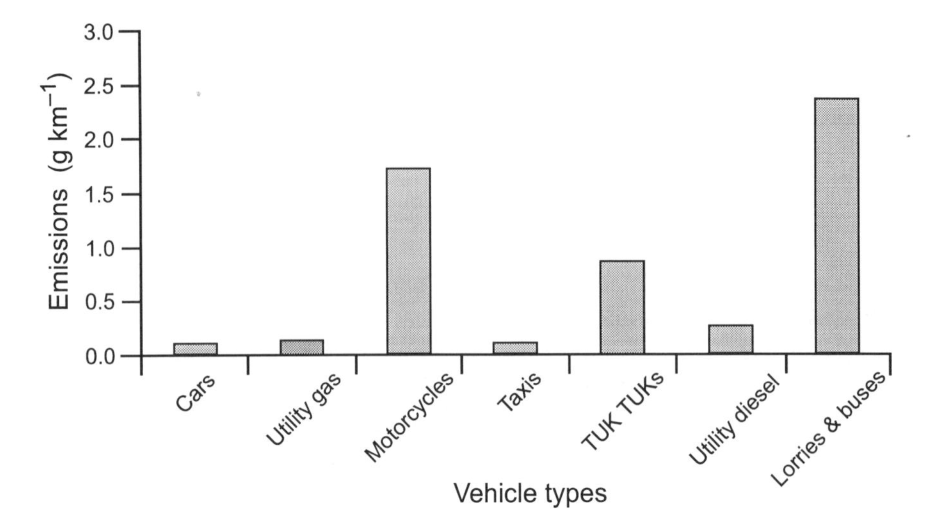

Figure 5.3 Emissions of organic suspended particulate matter by different types of vehicles in Bangkok, 1994 (Data provided by the Thai Ministry of Environment, Air Pollution Department)

oxidation or three-way catalysts have been placed on all new petrol fuelled cars. In Japan and Canada, catalysts are also widely used to meet emission standards. Many cars in Austria, the Netherlands, Sweden and Germany are sold with these systems, and they are required on new cars in Australia and Switzerland. One of the unique advantages of catalysts is their ability to eliminate selectively some of the more unhealthy compounds in vehicle exhaust, such as aldehydes, and reactive and polycyclic hydrocarbons.

Special problems with two stroke engines

Many of the two-wheeled vehicles which are prevalent in many developing countries are powered by two stroke engines. Two stroke motorcycles are a major source of white smoke, HC and SPM emissions. For example, in Bangkok two stroke powered motorcycles are estimated to emit almost as much SPM per kilometre driven as heavy duty diesel lorries and buses (Figure 5.3). In general, most of the solutions used to reduce emissions from four stroke engines can be applied to two stroke engines. However, because much of the HC emitted comes from the lubricating oil, it deserves special attention. Sugiura and Kagaya (1977) have pointed out that visible smoke can be reduced by the use of engine oils containing poly-isobutylene, and also by leaner fuel oil ratios. The separated lubrication system, which brings about overall leaner fuel/oil ratios is, therefore, favourable for smoke reduction. Refining the fairly simple type of carburettors used would help reduce HC, CO and smoke emissions significantly. Even catalytic converters are technologically feasible for these types of engines (OECD 1988a). Since 1986 mopeds with catalysts have been available in Switzerland and in Austria, in order to comply with the stringent emissions standards in these countries (Laimboch and Landerl, 1990) and since 1992 motorcycles in Taiwan have been similarly equipped. Today, therefore, the historical problems of high smoke and unburned HC associated with two stroke engines are no longer technologically necessary. New technology promises to resolve these concerns. As examples, direct cylinder electronic fuel injection, electronic computer control, and catalytic exhaust conversion are now commonplace solutions (Wyczalek, 1991).

In addition, modern two stroke engines are emerging which are starting to demonstrate very low emissions, excellent fuel economy and low cost. Foremost among these is the Orbital Engine which has undergone testing by the US EPA and, on at least one model, appears close to achieving the very stringent Ultra Low Emission Vehicle Standards of the State of California. These engines can therefor be cleaner and more fuel efficient than four-strokes.

5.4.2 Diesel fuelled vehicles

While the major technical problems associated with reducing emissions from petrol cars are being solved, it is apparent that reductions from these vehicles alone are not sufficient to solve air pollution problems in many areas. Diesel lorries and buses have also been receiving increased attention because they are significant sources of particulate matter and NO_X. With respect to other compounds, emissions from diesel vehicles are usually much lower than those of petrol driven vehicles. However, uncontrolled diesel engines emit objectionable exhaust odours which are a frequent source of complaints from the public. The USA has adopted standards for these vehicles which will foster technological developments similar to those which have already occurred for petrol cars. These developments are of great interest to other countries.

Smoke emissions from diesel engines are composed mainly of unburned carbon particles from the fuel and they are usually produced when there is an excess amount of fuel available for combustion. This condition is most likely to occur under high engine load conditions, such as during acceleration and when the engine is pulling, and needs additional fuel for power. Furthermore, the common maintenance error of failing to clean or replace a dirty air cleaner may produce high smoke emissions because the dirty filter can restrict available air to the engine, resulting in a lower than optimum air–fuel mixture. The operation of vehicles can also affect smoke emissions which, in diesel engines, can be minimised by selection of the proper transmission gear to keep the engine operating at the most efficient speeds. Moderate acceleration, less pronounced changes in speed, less frequent acceleration, more constant speed on highways, and reduced speed for hill climbing, also minimise smoke emissions.

The basic approaches to diesel engine emission control fall into three major categories:

- Engine modifications, including combustion chamber configuration and design, fuel injection timing and pattern, turbo charging and exhaust gas re-circulation.
- Exhaust after-treatment, including traps, trap oxidisers and catalysts.
- Fuel modifications, including control of fuel properties, fuel additives, and alternative fuels.

Control techniques for NO_X are being phased into diesel vehicles, including variable injection timing and pressure, charge cooling and exhaust gas recirculation. Retarding injection timing, a well known method of reducing NO_X formation, can lead to increases in fuel consumption and particulate and HC emissions. These problems can be mitigated by varying the injection timing with engine load or speed. High pressure injection can also reduce these problems; if coupled with electronic controls, it appears that NO_X emissions could be reduced significantly with a simultaneous improvement in fuel economy (although not as great as could occur if electronics were added without any emission requirements).

With relatively lenient particulate standards, engine modifications are generally sufficient to lower exhaust emission levels. Such modifications include changes in combustion chamber design, fuel injection timing and spray pattern, turbocharging, and the use of exhaust gas recirculation. Further particulate controls appear possible through the greater use of electronically controlled fuel injection which is currently under rapid development. Using such a system, signals (proportional to fuel rate and piston advance position) are measured by sensors and are electronically processed by the electronic control system to determine the optimum fuel rate and timing.

One of the important factors in diesel emission control design is that NO_X, particulate and HC emissions are closely interdependent; for example, retarded timing within certain ranges decreases NO_X and particulate matter but can lead to increases in HC. As technology has advanced, these potential trade-offs have diminished. Certain new engine designs (combustion chamber modifications, electronically controlled fuel injection, etc.) for example, have resulted in simultaneous reduction in HC, particulate matter and NO_X emissions. Exhaust after-treatment methods include traps, trap oxidisers and catalysts. Trap oxidiser prototype systems have shown themselves capable of 70–90 per cent reductions in exhaust particulate emission rates and, with proper regeneration, they could have the ability to achieve these rates for high mileage. Basically, all methods rely on trapping a major portion of the exhaust particles and consuming them before they accumulate sufficiently to saturate the filter and cause fuel economy, performance or other problems.

Recent studies have indicated that vehicles and engines designed for use off-road can be important sources of emissions. Such engines include equipment for lawns and gardens, airport servicing, recreation, light commercial purposes, industry, construction, agriculture, logging, and also commercial marine vessels. Both the US EPA and the California Air Resources Board (CARB) have developed standards for some of these sources. In addition, the European Union (EU) and the UN Economic Commission for Europe (UN ECE) are developing requirements for controls (see section 5.5).

5.5 World-wide progress in lowering vehicle emissions

Advances in automotive technologies have made it possible to lower emissions from motor vehicles dramatically. Countries around the world have been taking increasing advantage of them. Initial crankcase HC controls were first introduced in the early 1960s followed by exhaust CO and HC standards later that decade. By the early to mid-1970s, most major industrial countries had initiated some level of vehicle pollution control programme. For a variety of reasons, such as differing types and degrees of air pollution problems, varying vehicle characteristics, economic conditions, etc. the emission control approaches differ significantly between countries.

Progress in lowering emissions in petrol vehicles, heavy duty lorries, buses and other diesel powered vehicles is discussed in the following sections for developed countries and for countries in the process of industrialisation.

5.5.1 Petrol vehicles

Japan and the USA

During the mid- to late 1970s, advanced technologies were introduced to most new cars in the USA and Japan. These technologies resulted from a conscious decision to "force" the development of new approaches and were able to reduce CO, HC and NO_X emissions dramatically beyond those of previous systems. As knowledge of these technological developments on cars spread, and as the adverse effects of motor vehicle pollution became more

Table 5.2 Emission trends for passenger cars in the USA, 1970–90

Year		CO (10^6 t a^{-1})	HC (10^6 t a^{-1})	NO_x (10^6 t a^{-1})
1970	Actual	67.9	8.87	4.36
1990	Actual	26.9	2.65	2.34
1990	Potential	112.0	14.60	7.20

Source: US EPA, 1993

widely recognised, more and more people across the globe began demanding the use of these systems in their own countries.

The USA programme has combined many elements including the introduction of unleaded petrol, tight standards for new vehicles, in-use vehicle inspection and maintenance efforts and, most recently, the use of reformulated and low volatility petrol. As a result, between 1975 and 1995, the emission rate for on-highway cars in the USA declined dramatically. As newer vehicles equipped with advanced emission controls have replaced older, higher polluting ones, there has been a clear downward trend in emissions of all three pollutants. This is especially encouraging in the light of the continued rapid growth in vehicles and vehicle miles travelled by cars during this same period; in 1990 there were 50 million more cars on USA highways than there were in 1970. If emissions per mile had not been reduced, passenger cars in 1990 would have emitted 65 per cent more CO, HC and NO_X than they did in 1970. As illustrated in Table 5.2, instead of passenger car CO having been reduced from 68×10^6 t to 27×10^6 t, these emissions would have increased to 112×10^6 t. In Table 5.2 "Actual" denotes the actual emission trends, while "Potential" denotes what would have occurred if pollution controls not been put on cars over the period 1970–90.

Figure 5.4 illustrates auto-emission reductions to date, i.e. 61 per cent for CO, 70 per cent for non-methane hydrocarbons (NMHC) and 47 per cent for NO_X. Lead emissions from all highway vehicles have also been reduced dramatically; between 1970 and 1993, highway vehicle lead emissions declined from 171,960 to 1,380 tons (189,550 to 1,521 tonnes). This example shows that adoption of a strong motor vehicle pollution control programme can be very effective in reducing vehicle emissions. Another example is the experience in Southern California's Los Angeles Basin, which has had the most aggressive motor vehicle pollution control programme in the world over the past 40 years (Hall *et al.*, 1995). From 1955 to 1993, peak O_3 concentrations

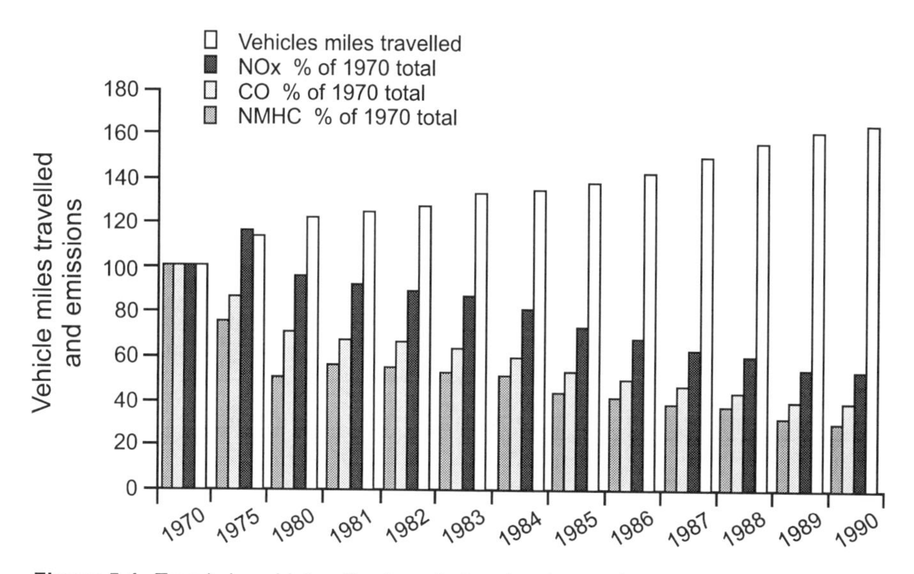

Figure 5.4 Trends in vehicle miles travelled and emissons (as a percentage of the 1970 total) from cars in the USA (After US EPA, 1993)

were reduced by 50 per cent. The number of days on which Federal O_3 standards were exceeded fell by 50 per cent from the 1976–78 time interval to the 1991–93 interval. Furthermore, the average annual number of days exceeding the Federal CO standard fell from 30 to 4.3 during this same period and Pb levels are now 98 per cent lower than in the early 1970s. Most remarkably, this achievement occurred while the regional economy exceeded the national economy in total job growth, manufacturing job growth, wage levels and average household income. In short, a strong focus on environmental protection is not incompatible with strong economic development, but it also seems to be mutually reinforcing.

Western Europe

During the mid-1980s, Austria, the Netherlands and the former Federal Republic of Germany adopted innovative economic incentive approaches to encourage the purchase of low pollution vehicles. Since then, Australia, Canada, Finland, Austria, Norway, Sweden, Denmark and Switzerland have all decided to adopt mandatory requirements. Within the EU, state of the art standards became mandatory in 1992/93 and a further tightening of standards is being phased in.

After extensive study (involving a fundamental auto industry–oil industry–EU Commission tripartite collaboration) followed by intense

Table 5.3 European Commission proposals for emission standards for petrol fuelled vehicles

Year	Standard	CO (g km^{-1})	HC (g km^{-1})	NO_x (g km^{-1})
1996/1997	Directive 94/12	3.2	0.341	0.252
2000 (Step 3)	Scenario 2	2.3	0.200	0.150
2005 (Step 4)	Scenario 4	1.0	0.100	0.080
UBA proposal		2.2	0.0675	0.140

UBA Umweltbundesampt (Federal Environmental Agency, Germany)

Sources: EU Council of Ministers, 1996; Walsh, 1996

internal debate reflecting a variety of interests, the European Commission finally released, in 1996, its long-awaited proposal for reducing automobile emissions in the year 2000 and beyond. Although it was delayed about 18 months from the original deadline set by the Council of Ministers (i.e. end December 1994), the proposal appeared in many ways to have been worth the wait. It not only established the link between vehicle and fuel standards and public health needs, but it also introduced onboard diagnostics and, for the first time in the EU, an in-use recall liability for vehicle manufacturers. In spite of substantial resistance by some companies, it also included additional targets for vehicles in the year 2005, which can serve as the basis for fiscal incentives by individual member states and will be reviewed again in 1998 to evaluate their technological feasibility. Additional fuel requirements will also be reconsidered at that time.

The emission standards proposed for petrol-fuelled light duty vehicles are summarised in Table 5.3. For comparison purposes, current Step 2 (Directive 94/12) and the proposals of the Federal Environmental Agency of Germany (Umweltbundesamt, UBA) of April 1995 are also included.

The test cycle will be modified by the Motor Vehicle Emission Group of the European Commission (MVEG) to eliminate the 40 second warm-up before measurement of emission begins. The base case Directive 94/12 shown in Table 5.3 for petrol fuelled vehicles (and below in Table 5.6 for diesel fuelled vehicles) has been adjusted to account for this test procedure revision, as previously agreed by MVEG. The Step 4 values for the year 2005 shown above reflect the Commission's judgement of standards which would be slightly more stringent than best available technology for the year 2000.

Lead will be banned from petrol as from the year 2000, along with a number of other improvements in fuel composition summarised in Table 5.4.

Table 5.4 European Commission proposals for petrol composition by the year 2000

Parameter	Average today	Proposed average	Proposed maximum
Reid vapour pressure (RVP) (kPa) — summer	68	58	60
Aromatics (% v)	40	37	45
Benzene (% v)	2.3	1.6	2
Sulphur (ppm)	300	150	200

Sources: EU Council of Ministers, 1996; Walsh, 1996

Rapidly industrialising countries

Even rapidly industrialising, developing countries such as Brazil, Mexico, Taiwan and South Korea have adopted stringent emission regulations. More recently, similar requirements have been introduced by Chile, Hong Kong, Thailand and Singapore. Adoption of current state-of-the-art emission standards has been demonstrated to lower emissions substantially.

5.5.2 Heavy duty lorries, buses and other diesels

As a result of the various problems associated with diesel smoke and particulates, control programmes have been under development for many years. This section reviews the history of these programmes to date. In general, the initial focus of such programmes was smoke control because smoke was clearly visible and considered to be a nuisance. As the evidence for serious health and environmental risks associated with diesel exhausts has grown in recent years, more attention has focused on control of the particles themselves, particularly for health reasons. Although smoke standards lower visible smoke, they are not as effective at lowering particulate emissions as standards for particulates. There is some, but not complete, correlation between visible smoke and particulates. Smoke standards are designed to address problems of visibility and particulate standards are designed to achieve acceptable ambient particulate levels.

USA

Emission control requirements in the USA for smoke from engines used in heavy duty lorries and buses were first implemented for vehicles manufactured in 1970. These opacity standards were specified in terms of per cent of light allowed to be blocked by the smoke in the diesel exhaust (as determined by a light extinction meter). Heavy duty diesel engines produced during the years 1970 through to 1973 were allowed a light extinction of 40 per cent during the acceleration phase of the certification test and 20 per cent during the lugging (pulling) portion; vehicles produced in 1974 and later are subject to smoke opacity standards of 20 per cent during acceleration, 15 per cent during lugging and 50 per cent at maximum power.

The first diesel exhaust particulate standards in the world were established for cars and light lorries by the US EPA on 5 March, 1980. Standards of 0.6 grams per mile (0.37 g km^{-1}) were set for all cars and light lorries starting with those manufactured in 1982 and dropping to 0.2 grams per mile (0.12 g km^{-1}) and 0.26 grams per mile (0.16 g km^{-1}) for cars produced in 1985 and light lorries, respectively. In early 1984, the US EPA delayed the second phase of the standards from 1985 to cars manufactured in 1987 in order to provide more time for manufacturers to comply. Almost simultaneously, California decided to adopt its own diesel particulate standards of 0.4 grams per mile (0.25 g km^{-1}) in 1985, 0.2 grams per mile (0.12 g km^{-1}) in 1986 and 1987, and 0.08 grams per mile (0.05 g km^{-1}) in 1989.

Particulate standards for heavy duty diesel engines were promulgated by the US EPA in March, 1985. Standards of 0.60 grams per brake-horsepower-hour (g/bhph) (equivalent to 0.80 grams per kilowatt hour (g/kWh)) were adopted for vehicles manufactured from 1988 through to 1990, 0.25 g/bhph (0.34 g/kWh) for 1991 through to 1993 and 0.10 g/bhph (0.13 g/kWh) for 1994 and later years. Because of the special need for bus control in urban areas, the 0.10 g/bhph (0.13 g/kWh) standard for these vehicles went into effect in 1993, and was then reduced by an additional 50 per cent one year later.

Subsequently, the US EPA revised the 0.26 grams per mile diesel particulate standard for certain light duty lorries. Light duty diesel trucks (LDDTs) with a loaded vehicle weight of 3,751 pounds (1,701.5 kg) or greater, otherwise known as LDDT2s, were required to meet a 0.50 grams per mile standard for 1987 and 0.45 grams per mile level for 1988–90. For lorries produced in 1991 and later model years the standard was tightened to 0.13 grams per mile. It is worth noting that the driving force behind the control of particulate emissions from vehicles in the USA was the adverse impacts of particles on mortality and morbidity. While concerns have existed regarding potential carcinogenic effects from exposure to diesel exhaust, this may not have been the basis for regulation of these emissions.

The most important development in the USA in relation to regulating emissions from diesel vehicles and engines has been the move towards more stringent health-based air quality standards for O_3 and SPM. On 27

November 1996, the US EPA announced its proposals to change the standards for both pollutants (Walsh, 1997). For O_3, it was proposed to lower the current standard from 0.12 ppm (240 $\mu g\ m^{-3}$) to 0.08 ppm (160 $\mu g\ m^{-3}$). In addition, attainment of the standard would be based on 8-hour averages instead of the present 1-hour averages. Based on the studies of health effects of SPM performed in the USA and described in Chapter 2, the US EPA has also proposed to revise the current primary PM_{10} standards by adding two new primary $PM_{2.5}$ standards set at 15 $\mu g\ m^{-3}$, annual mean, and 50 $\mu g\ m^{-3}$, 24-hour average. It also proposed to retain the current annual primary PM_{10} standard of 50 $\mu g\ m^{-3}$.

As understanding of the photochemical phenomena related to O_3 has developed, NO_X control options have received increasing attention. Control of NO_X has emerged as the primary strategy, particularly in addressing regional-scale O_3 problems. In recent years, the US EPA has conducted modelling studies covering the eastern half of the USA. These studies have reinforced the understanding that regional-scale control of NO_X emissions will be essential for reducing the levels of transported O_3 in large areas of the Northeast, Southeast and Midwest (US EPA, 1997). The US EPA has adopted a new combined emission standard for NO_X and HC for vehicle model year 2004 and later heavy duty diesel engines used in lorries and buses. The new standard represents a 50 per cent reduction in NO_X from the 1998 and later model year NO_X standard. Details of the new requirements are:

- Engine manufacturers will have a choice of certifying heavy duty diesel engines to either of two optional sets of standards: 2.4 g/bhph NMHC+NO_X, or 2.5 g/bhph NMHC+NO_X with a limit of 0.5 g/bhph on NMHC.
- All emissions standards other than NMHC and NO_X applying to 1998 and later model year heavy duty engines continue at their 1998 levels.

The US EPA is also providing for a 1999 review of the standards set in this rule. This review will reassess whether the standards are still appropriate under the Clean Air Act, together with the need for, and technical and economical feasibility of, the standards, based on information available in 1999. The US EPA believes that the 2004 model year standards are technologically feasible without any changes to diesel fuel. As part of the 1999 review, it will evaluate whether diesel fuel improvements are needed for the standards to be appropriate in the year 2004, taking into consideration any new information. If the US EPA finds that diesel fuel changes are needed to meet the standards, and if it believes that such changes would be a cost-effective method for reducing emissions, then the US EPA will address the potential for fuel improvements through a separate "Rulemaking".

The US EPA is finalising provisions to enhance the control of emissions from in-use vehicles subject to the new model year 2004 standards. The in-use provisions include:

- Revisions of existing regulations, including useful life, emission-related maintenance and emission defect and performance warranties.
- New provisions regarding maintenance and repair of emission controls after the end of the useful life, including manufacturer requirements and engine rebuild provisions.

All of the changes to the regulations will be effective beginning with the 2004 model year.

Canada

In March 1985, in parallel with a significant tightening of gaseous emission standards, Canada adopted the USA particulate standards for cars and light lorries (0.2 and 0.26 grams per mile, respectively) and brought them into effect for vehicles manufactured in 1988. Since then, Canada have also decided to adopt the USA standards for heavy duty vehicles for 1988 and has subsequently linked its standards to those in the USA such that manufacturers supply vehicles meeting the same emission standards to each market.

Japan

In Japan, smoke standards have applied to both new vehicles and those in-use since 1972 and 1975, respectively; the maximum permissible limits in each case are 50 per cent opacity. The new vehicle standard is however the more stringent because smoke is measured at full load, while in-use vehicles are required to meet standards under the less severe no-load acceleration test. The Japanese Environment Protection Agency is also moving forward with its regulation of diesel vehicles. The long-term targets identified in 1989 will be phased-in over the period from 1997–99 as shown in Table 5.5.

European Community

Road vehicles. Smoke limits similar to those described above in the USA and Japan have been in effect in Europe for many years. Exhaust smoke levels are currently recommended by EC Regulation 24 (equivalent to EEC Directive 72/306). Measurements are taken using light absorption apparatus. However, recognising that these requirements are not adequate, the European Community Environment Ministers decided, in December 1987, to adopt a particulate standard for light duty diesel vehicles. First stage requirements have now been introduced and will be gradually tightened throughout the

Table 5.5 Japanese standards for diesel fuelled vehicles

Vehicle category	NO_x	PM	Year of implementation
	($g\ km^{-1}$)	($g\ km^{-1}$)	
GVW < 1.7 t	0.4	0.08	1997
1.7 t < GVW < 2.5 t (M)	0.7	0.09	1997
1.7 t < GVW < 2.5 t (A)	0.7	0.09	1998
	(g/kWh)	(g/kWh)	
2.5 t < GVW < 3.5 t	4.5	0.25	1997
3.5 t < GVW < 12 t	4.5	0.25	1998
GVW > 12 t	4.5	0.25	1999

GVW Gross vehicle weight

Source: T. Ijima, Environment Agency, Air Quality Bureau, Japan

Table 5.6 Standards proposed by the European Commission for light duty and diesel fuelled vehicles

Year	CO ($g\ km^{-1}$)	HC ($g\ km^{-1}$)	HC and NO_x ($g\ km^{-1}$)	NO_x ($g\ km^{-1}$)	PM ($g\ km^{-1}$)
1996/1997 Directive 94/12	1.06		0.71/0.91[1]	0.63/0.81[1]	0.08/0.10[1]
2000 (Step 3) Scenario 2	0.64		0.56	0.50	0.05
2005 (Step 4) Scenario 3	0.50		0.30	0.25	0.025
UBA proposal (2000)	1.0	0.05		0.40	0.050
UBA proposal (2003)	1.0	0.05		0.14	0.025

1 Direct injection engine
UBA Umweltbundesampt (Federal Environmental Agency, Germany)

Source: EU Council of Ministers, 1996; Walsh, 1996

1990s. The standards proposed for light duty and diesel-fuelled vehicles are summarised in Table 5.6.

On 18 June 1996, the European Commission released its Step 3 and Step 4 proposals for reducing automobile emissions in the year 2000 and beyond. This proposal established the link between vehicle and fuel standards and public health needs and also, for the first time in the EU, introduced an in-use recall liability for vehicle manufacturers. The simultaneous proposal of Step 3 and Step 4 requirements with the Step 4 levels to serve as a basis for fiscal incentives by individual member states, was particularly innovative. The light duty vehicle standards proposed and adopted over the past decade are summarised in Table 5.7. The test cycle will be modified for Steps 3 and 4 to eliminate the 40 second warm-up before emission measurements begin (Step 2 is shown twice to illustrate the effect of this test procedure revision).

Table 5.7 Proposed EU light duty vehicle exhaust emission standards

Step	Vehicle type	CO ($g\ km^{-1}$)	HC and NO_x ($g\ km^{-1}$)	HC ($g\ km^{-1}$)	NO_x ($g\ km^{-1}$)	PM ($g\ km^{-1}$)
1	All [1]	3.16	1.13			0.18 [2]
2	Petrol	2.0	0.5			
2	Diesel	1.0	0.7			0.08
2	Diesel DI [3]	1.0	0.9			0.10
2 Mod. Test	Petrol	3.2			0.252	
2 Mod. Test	Diesel	1.06			0.63	0.08
2 Mod. Test	Diesel DI [3]	1.06			0.81	0.10
3	Petrol	2.3		0.2	0.15	
3	Diesel	0.64	0.56		0.5	0.05
4	Petrol	1.0		0.1	0.08	
4	Diesel	0.5	0.3		0.25	0.025

[1] Conformity of production
[2] Diesel only
[3] Direct injection diesel, 3 years only

Sources: EU Council of Ministers, 1996; Walsh, 1997; VDI Nachrichten, 10 July 1998

In December 1995, the Council of Ministers also reached a common position regarding tighter standards for light lorries. These standards are summarised in Table 5.8. As part of the overall effort that produced these standards, the Commission is committed to proposing new standards for light commercial vehicles in the near future. Preliminary levels under consideration are summarised in Table 5.9.

In a first step towards reducing emissions from heavy duty vehicles, the European Commission adopted Directive 88/77/EEC, which applied to vehicles larger than 3.5 t as of 1 October 1990. Under this Directive, standards of 11.2 g/kWh CO, 2.4 g/kWh HC and 14.4 g/kWh NO_X, as measured on an engine test procedure based on the former USA 13-mode test, were promulgated. At 1 October 1991 meeting of the Council of Environment Ministers, final agreement was reached on a heavy duty diesel engine Directive which put in place a long-term strategy for reducing diesel emissions. The unanimous decision by the Environment Ministers applied to particulate and gaseous emissions as summarised in Table 5.10. The Directive took effect in two stages. New vehicle models were required to conform to Step 1 by 1 July 1992 and all new vehicles by 1 October 1993. The second stage began 1 October 1995, for new vehicle models, and 1 October 1996, for all vehicles. Fuel with 0.05 per cent sulphur by weight will have to be available for the second stage and manufacturers are allowed to certify their engines using this new fuel immediately. The sulphur content in diesel fuel was

Table 5.8 EU light lorry exhaust emission standards

Weight of vehicle	Fuel	CO (g km^{-1})	HC and NO_x (g km^{-1})	PM[1] (g km^{-1})	Date of application
Class 1	All	2.72	0.97	0.14	1 Oct. 1993
(< 1,250 kg)					1 Oct. 1994
	Petrol	2.2	0.5		1 Oct. 1997
	Diesel IDI	1.0	0.7	0.08	1 Oct. 1997
	Diesel DI	1.0	0.9	0.10	1 Oct. 1997
	Diesel DI	1.0	0.7	0.08	1 Oct. 1999
Class 2	All	5.17	1.40	0.19	1 Oct. 1993
(1,251–1,700 kg)					1 Oct. 1994
	Petrol	4.0	0.6		1 Oct. 1998
	Diesel IDI	1.25	1.0	0.12	1 Oct. 1998
	Diesel DI	1.25	1.3	0.14	1 Oct. 1998
	Diesel DI	1.25	1.1	0.14	1 Oct. 1999
Class 3	All	6.9	1.7	0.25	1 Oct. 1993
(< 1,700 kg)	Petrol	5.0	0.7		1 Oct. 1994
	Diesel IDI	1.5	1.2	0.17	1 Oct. 1998
	Diesel DI	1.5	1.6	0.20	1 Oct. 1998
	Diesel DI	1.5	1.3	0.20	1 Oct. 1999

IDI Indirect injection
DI Direct injection
[1] Diesel vehicles only

Source: EU Council of Ministers, 1996; Walsh, 1997

Table 5.9 EU preliminary proposed standards for exhaust emissions from light lorries

Category	Class	Reference weight (kg)	CO (g km^{-1}) Petrol	CO (g km^{-1}) Diesel	HC and NO_x (g km^{-1}) Petrol	HC and NO_x (g km^{-1}) Diesel	PM (g km^{-1}) Diesel
N1	I	< 1,250	2.29	0.64	0.36	0.50	0.05
N1	II	1,250–1,700	4.17	0.80	0.43	0.72	0.08
N1	III	> 1,700	5.22	0.95	0.50	0.85	0.11

Source: EU Council of Ministers, 1996; Walsh, 1997

reduced to no more than 0.2 per cent (weight) beginning 1 October 1994, and was reduced to 0.05 per cent (weight) starting 1 October 1996. As of 1 October 1995, there had to be a balanced distribution of available diesel fuel, so that at least 25 per cent of diesel fuel in each member country had a sulphur content of not more than 0.05 per cent (weight). Further improvements, summarised below, are under discussion:

Pollutant	Present average	Proposed average	Proposed maximum
Polyaromatics (% v)	9	6	11
Sulphur (ppm)	450	300	350

In 1997, the European Commission issued its proposal relating to Step 3 (Euro 3) and Step 4 (Euro 4) stages for heavy duty vehicles. The proposed Euro 3 limits are given in Table 5.11. These limits are intended to achieve a 30 per cent reduction from Euro 2, in accordance with the recommendations resulting from Auto/Oil 1 (see Peake, 1997). They will go into effect on 1 October 2000 for new types of vehicles and 1 October 2001 for registration, sale and entry into service.

In order to allow time for the potential development of a new world-wide harmonised test cycle to gain a better understanding of heavy duty engine control technologies, no Euro 4 limits have been proposed. Nevertheless, an emission reduction target of up to 40 per cent, compared with Euro 3, is being considered. It is intended that the Commission will make proposals for Euro 4 limits by 31 December 1999, taking into account:

- The results of Auto/Oil 2 (see Peake, 1997).
- Developments in respect of emission control technology, including the interdependence with fuel quality.
- Development of the world-wide harmonised test cycle.
- Development of on-board diagnostics (OBD) for heavy duty engines.
- The needs for specific durability provisions for diesel and gas engines.

Until Euro 4 is actually proposed, fiscal incentives may only apply to the Euro 3 limits.

Off road vehicles and engines. In June 1996, the Environment Ministers of the European Union reached a “common position” on the first European legislation to regulate emissions from non-road, mobile equipment such as road construction tractors, graders, loaders, compressors, and some agricultural devices. The primary goal of the legislation is to help the European Union reduce smog and unhealthy or acid rain-inducing chemicals such as oxides of nitrogen, CO and particulate matter. The legislation also will help the EU to meet its legal requirements under the UN ECE protocol that calls for reductions in oxides of nitrogen and VOCs. The new requirements are expected to go into force by 1998. They will be phased in over two stages and will set restrictions on emissions based on the size of the engines. However, the directive must first be approved by the European Parliament because it falls under Article 100a of the Maastricht treaty. Article 100a legislation

Table 5.10 Heavy duty diesel requirements in the European Union

	CO (g/kWh)	HC (g/kWh)	NO_X (g/kWh)	PM (g/kWh)
EC Step 1 (Euro 1)	Type approval (TA)			
	4.5	1.1	8.0	0.63 (< 85 kW) 0.36 (> 85 kW)
	Conformity of production (COP)			
	(4.9)	(1.23)	(9.0)	(0.7) (< 85 kW) (0.4) (> 85 kW)
EC Step 2 (Euro 2)	TA and COP			
	4.0	1.1	7.0	0.15

Source: EU Council of Ministers, 1996; Walsh, 1996

Table 5.11 EC Step 3 (Euro 3) heavy duty diesel requirements

Technology	Test	CO (g/kWh)	HC (g/kWh)	NO_X (g/kWh)	PM (g/kWh)		Smoke (m-1)
Conventional diesel	ESC and ELR (OICA)	2.1	0.66	5.0	0.10	0.16 [1]	0.8
Advanced diesel	ETC (FIGE)	5.45	0.85	5.0	0.16	0.25 [1]	n/a
Gas	ETC (FIGE)	5.45	0.85	5.0	n/a		n/a

ESC Environmental Systems Corporation, Knoxville, Tennessee
ETC Environmental Technology Centre, Ottawa, Ontario
OICA Organization Internationale des Constructeurs d'Automobiles
FIGE Forschungsinstitute Geräusche und Erschütterungen, Herzogenrath, Germany

n/a Not applicable
1 For engines having a swept volume of less than 0.7 litres per cylinder and a rated power speed of more than 3,000 revolutions per minute

Source: European Council of Ministers, 1997; Walsh, 1998

covers laws dealing with internal market measures and therefore gives co-decision procedures to the European Parliament.

Studies carried out by the European Commission indicate that once fully implemented, the proposals will lead to reductions in emissions from non-road mobile machinery of up to 42 per cent in the case of NO_X and 67 per cent in the case of particulate matter. According to the Commission, the total annual costs for technically upgrading the engines were estimated to be in the range of 31 million ECU (US$ 38 million) for Stage I and 125 million ECU (US$ 153 million) for Stage II. This would result in an increase in retail prices

Table 5.12 Stage I emission limits as proposed by the European Commission

Net power (kW)	CO (g/kWh)	HC (g/kWh)	NO_x (g/kWh)	Particulates (g/kWh)	Date of implementation
130 < P < 560	50	13	92	54	30 September 1998
75 < P < 130	50	13	92	70	30 September 1998
37 < P < 75	65	13	92	85	31 March 1999

Source: EU Council of Ministers, 1996; Walsh, 1997

of about 1 per cent during Stage I and III to 8 per cent during Stage II. A third stage is included in the legislation, but the requirements will be decided at the end of the year 2000.

The common position on the new legislation was reached after member states, led by the UK, demanded that economic incentives originally proposed by the Commission should be dropped. The European Commission had argued that tax incentives would push industry to adopt stricter environment standards as outlined in the second stage of the directive. However, industry representatives, including those from the Brussels-based Association of European Manufacturers of Internal Combustion Engines, claimed that tax incentives would confuse the market, thereby disrupting the EU's internal market.

The machinery covered by the legislation includes industrial drilling rigs, compressors, and construction equipment including wheel loaders, bulldozers, off-highway lorries, highway excavators and fork lift trucks, road maintenance equipment, snowplough equipment, ground support equipment in airports, aerial lifts, and mobile cranes.

Stage I and Stage II emissions must not exceed the levels shown in Tables 5.12 and 5.13 respectively. These emission limits are exhaust limits and must be achieved before any exhaust after-treatment device. Most agricultural and forestry machinery will be exempt from the legislation.

Future plans. In the view of many Europeans, regulations adopted to date in Europe are not considered sufficient to protect public health and the environment. For example, a UK report noted that projections of future emissions, taking into account anticipated traffic growth and stricter emission controls, indicated that an increased market penetration of diesel cars at the expense of three way catalyst petrol cars would, on balance, have a deleterious effect on urban air quality (Quality of Urban Air Review Group, 1993). The report also highlighted that the black smoke component of particulate matter, which in

Table 5.13 Stage II emission limits proposed by the European Commission

Net power (kW)	CO (g/kWh)	HC (g/kWh)	NO_X (g/kWh)	Particulates (g/kWh)	Date of implementation
130 < P < 560	35	10	60	2	31 December 2000
75 < P < 130	35	10	60	3	31 December 2001
37 < P < 75	50	13	70	4	31 December 2002
18 < P < 37	55	15	80	8	31 December 2003

Source: EU Council of Ministers, 1996; Walsh, 1997

most urban areas was due almost entirely to diesel emissions, was responsible for the soiling of buildings. Fine particulate matter was also associated with reduced visibility and had been linked to a range of adverse health effects, for which there was a continuous dose-response relationship but without a no-effect threshold. Consequently, any slowing in progress towards lower particulate matter concentrations would be highly undesirable (Quality of Urban Air Review Group, 1993).

The question of more stringent requirements for diesel vehicles is therefore under active consideration by the European Commission for both light and heavy commercial vehicles and for passenger cars. With heavy duty engines, the appropriate test procedure is also a subject of discussion for the possible Stage III standards. Standards currently being discussed for light duty vehicles are 0.04 g km^{-1} for particulate matter and 0.5 g km^{-1} for NO_X; for heavy duty vehicles, NO_X and particulate levels of 5.0 and 0.10 g/kWh respectively, are being discussed. These requirements will challenge engine manufacturers, fuel providers and after-treatment system suppliers, and yet may not be completely adequate to protect public health. The Volvo car manufacturing company recently indicated the likely prospect of "Euro 4" and "Euro 5" emission levels of 3.0 and 2.0 g/kWh NO_X, respectively, being phased in after the turn of the century; in both cases the particulate level would remain 0.1 g/kWh (Sterner, 1994).

5.6 Future prospects for vehicle trends

The development of petroleum-powered motor vehicles has truly revolutionised society over the past century. The benefits of increased personal mobility and access to goods and services, previously beyond the grasp of individuals, cannot be denied. However, the relentless growth in motor vehicle use has adverse effects that many have been slow to acknowledge, including a broad range of public health and environmental effects. The environmental damage caused by motor vehicle emissions is now widely accepted and is increasing on a global scale. The cars, lorries and buses that make life better in so many ways emit more than 800 × 10^6 t of carbon a year. From their exhaust pipes comes virtually all of the CO in the air of the world's cities. Less directly, motor vehicles are responsible for much of the O_3 and smog, and they play a significant role in stratographic O_3 depletion. All of these pollutants contribute directly or indirectly to global warming.

The growth in demand for motorised travel is well understood. As urban areas populate and expand, land which is generally at the edges of the urban area and previously considered unsuitable for development, is developed. Also, the high cost of housing in the city centres forces people to pursue affordable housing in the suburban areas. As the distance of these residential locations from the city centre or other sub-centres increases, so does the need for motorised travel. Motorised travel often in private vehicles supplants traditional modes of travel, such as walking, various bicycle forms, water travel and even mass transit (public mass transport). The need for private vehicles is reinforced as declining population and employment densities with distances from urban centres reduce the economic viability of mass transit.

The evolution of the form of urban areas is influenced by growth in income and the accompanying increases in the acquisitions of private motor vehicles and changes in travel habits. As incomes rise, an increasing proportion of journeys are made first by motorcycles and then, as income increases further, by private cars. The trend towards private motorisation is also influenced by public policy towards land use, housing and transportation infrastructure. Whereas the proportion of middle and upper income households in developing and newly industrialised countries able to afford cars and motorcycles is lower than in industrialised nations, the number of private vehicles becomes very large with the growth of middle and upper income groups in megacities. The number of vehicles and levels of congestion are comparable with, or exceed that of, major cities in industrialised countries. With the increase in motorised travel and congestion come increases in energy use, emissions and air pollution (TDRI, 1990).

Projections of population trends for the future in most rapidly industrialising countries indicate both rapid population growth and increasing urbanisation of that population. In short, these trends generally increase the geographical spread of cities, both large and small, increasing the need for motorised transport to carry out an increasing portion of daily activities. Furthermore, when coupled with expanding economies (as is increasingly the case in these countries), a greater proportion of the urban population can afford personal motorised transportation, starting with motor cycles and progressing, as soon as economically feasible, to cars.

Over the last 40 years, the global vehicle fleet has grown from under 50 million to more than 500 million, and there is every indication that this growth will continue. Over the next 20 years, the global fleet could double to 1,000 million. Unless transportation technology and planning are fundamentally transformed, emissions of greenhouse and other polluting gases from these vehicles will continue to increase, many relatively clean environments will deteriorate, and the few areas that have made progress will see some of their gains eroded.

The world-wide challenges that these problems pose for motor vehicle manufacturers and policy-makers are unprecedented. Nothing less than a revolution in technology and thinking, at least as profound as the initial mechanisation of transportation, is needed. Manufacturers will come under increasing pressure to produce petroleum-powered vehicles that are ever cleaner, safer, more reliable, and more fuel efficient. At the same time, they will need to develop new kinds of vehicles that will emit no pollution whatsoever. The amount of capital needed to accomplish these goals will be large and, making matters even more difficult, the pressures for these changes will arise not so much from traditional market forces but from public policies adopted in response to climate change and other threats.

While appropriate policies and technologies continue to develop, countries can benefit from the adoption of those that are currently available. Various steps can be taken to reduce air pollution emissions from motor vehicles. These include incentives to remove older, higher polluting vehicles from the road; tightening new vehicle emission standards for NO_X, VOCs, and CO; developing and using cleaner fuels with lower volatility and fewer toxic components; enhancing I/M programmes, including inspections of anti-tampering emission-control equipment; and extending the useful life of pollution-control equipment to 10 years or 100,000 miles rather than the current five years or 50,000 miles. The potential overall impact of tighter standards, enhanced I/M, and extended useful life is especially significant because it helps to ensure that the benefits of clean-air technology will persist for the full life of the vehicle.

Additional reductions in vehicular emissions can be achieved by reducing dependence on individual cars and lorries and by making greater use of van and car pools, buses, trolleys and trains. Improving urban traffic management by installing synchronised traffic lights, reducing on-street parking, switching to "smart" roads, banning truck unloading during the day, etc. can also improve transportation system efficiency (Office of Technology Assessment, 1989).

Providing efficient, convenient, and affordable public transportation alternatives world-wide would produce multiple benefits. When 40 persons get out of their cars and onto a bus for a 10 mile trip to work, the emission of some 50 to 75 pounds of carbon into the air is avoided. Greater use of public transportation would reduce congestion, cut fatalities and injuries from traffic accidents, and greatly improve air quality.

In summary, therefore, in coming years, under current policies many more citizens of developing countries can be expected to be living in larger and larger cities and driving more and more private vehicles. As a result, without intervention, the already serious pollution problems of today can be expected to get worse, exposing greater numbers of people to even higher levels of pollution for longer periods of time.

5.7 References

Auto/Oil 1990 *Initial Mass Exhaust Emissions Results from Reformulated Gasolines.* Air Quality Improvement Research Program, Technical Bulletin No. 1, Auto/Oil Programme, EU Task Force, London.

Auto/Oil 1991 *Effects of Fuel Sulfur Levels on Mass Exhaust Emissions.* Air Quality Improvement Research Program, Technical Bulletin No. 2, Auto/Oil Programme, EU Task Force, London.

Braddock, J.N. 1988 *Factors Influencing the Composition and Quantity of Passenger Car Refuelling Emissions–Part II.* SAE Paper No. 880712, Society of Automotive Engineers – The Engineering Society for Advancing Mobility on Land, Sea, Air and Space, Warrendale, Pennsylvania.

Colucci and Wise 1992 *What Is It and What Has It Learned? Auto/Oil Air Quality Improvement Research Program,* Presented at XXIV Fisita Congress London, England.

EU Council of Ministers 1996 EU Commission proposal for petrol and diesel fuelled vehicles. 18 June 1996.

Hall, J.V., Brajer, V., Lurmann, F.W. and Colome, S.D. 1995 *The Automobile, Air Pollution Regulations and the Economy of Southern California, 1965-1990.* Department of Economics, California State University, Fullerham, CA 92637.

Laimboch, P. and Landerl, L. 1990 *50cc Two-Stroke Engines for Mopeds, Chainsaws and Motorcycles with Catalysts.* SAE Paper No. 901598, Society of Automotive Engineers – The Engineering Society for Advancing Mobility on Land, Sea, Air and Space, Warrendale, Pennsylvania.

McArragher, J.S. 1988 *Evaporative Emissions from Modern European Vehicles and their Control.* SAE Paper No. 880315, Society of Automotive Engineers – The Engineering Society for Advancing Mobility on Land, Sea, Air and Space, Warrendale, Pennsylvania.

OECD 1988 *Transport and Environment*. Organisation for Economic Co-operation and Development, Paris.

Office of Technology Assessment 1989 Advanced vehicle/highway systems and urban traffic problems. Staff Paper. Science, Education, and Transportation Program, Office of Technology Assessment, US Congress.

Peake, S. 1997 *Vehicle Fuel Challenges Beyond 2000. Market Impacts of the EU's Auto Oil Programme*. Financial Times Automotive Publishing, London.

Quality of Urban Air Review Group 1993 *Diesel Vehicle Emissions and Urban Air Quality*. Second Report prepared at the request of the Department of the Environment, London.

Sterner, K. 1994 *Volvo Environmental Programme, Volvo Truck Corporation, Sweden.* Paper presented at the Symposium on Indian Automotive Technology, New Delhi, 5–9 September 1994.

Sugiura, Y. and Kagaya, T. 1977 *A Study of Visible Smoke Reduction from a Small Two-Stroke Engine Using Various Lubricants.* SAE Paper No. 770623, Society of Automotive Engineers – The Engineering Society for Advancing Mobility on Land, Sea, Air and Space, Warrendale, Pennsylvania.

TDRI 1990 *The 1990 TDRI Year End Conference, Industrializing Thailand And Its Impact On The Environment.* Research Report No. 7, Energy and Environment: Choosing The Right Mix, Thailand Development Research Institute, Bangkok.

US EPA 1987 *Draft Regulatory Impact Analysis. Control of Gasoline Volatility and Evaporative Hydrocarbon Emissions From New Motor Vehicles*. Office of Mobile Sources, United Sates Environmental Protection Agency, Washington, D.C.

US EPA 1993 *National Air Quality and Emission Trends Report 1993.* EPA 454/R-94-026, United Sates Environmental Protection Agency, Washington, D.C.

US EPA 1994 *Cost Effectiveness of Reformulated Gasolines.* United Sates Environmental Protection Agency, Washington, D.C.

US EPA 1997 *NPRM on Heavy Duty Engines*. United Sates Environmental Protection Agency, Washington, D.C.

Walsh, M. 1996 *Carlines.* Issue 96-4, July 1996, Arlington, VA.

Walsh, M. 1997 *Global Trends in Diesel Emission Control — A 1997 Update.* SAE Technical Paper Series No 970179, Society of Automotive Engineers – The Engineering Society for Advancing Mobility on Land, Sea, Air and Space, Warrendale, Pennsylvania.

Walsh, M. 1998 *Global Trends in Diesel Emission Control — A 1998 Update*. SAE Technical Paper Series No 980186, Society of Automotive Engineers – The Engineering Society for Advancing Mobility on Land, Sea, Air and Space, Warrendale, Pennsylvania.

Wyczalek, P. 1991 Two-Stroke Engine Technology in the 1990's. SAE 910663, Contained in SP-849, *Two Stroke Engine Design and Development*. Society of Automotive Engineers – The Engineering Society for Advancing Mobility on Land, Sea, Air and Space, Warrendale, Pennsylvania.

Chapter 6*

CASE STUDIES FROM CITIES AROUND THE WORLD

In most major cities of developing countries, there is a large and growing population of poorly maintained vehicles with few, if any, pollution controls. Many of these vehicles are powered by unusually dirty fuels. By the late 1980s, therefore, most of the cities were experiencing serious motor vehicle-related air pollution problems, often in addition to other serious environmental problems. In virtually every city for which data are available, CO, Pb and SPM are the primary pollutants causing the problems. Furthermore, vehicles contribute significant amounts of HC and NO_X emissions. These are frequently toxic and contribute to photochemical smog in cities with unfavourable meteorological conditions.

Quantifying the cost of air pollution is very difficult. According to the Hong Kong EPA (1989):

> *"Although it is impossible to be precise, hundreds of millions of dollars are spent every year on combating air pollution or paying to rectify its effects. This expenditure arises in many different ways, from the obvious, such as maintaining the EPDs (Environment Protection Department) air control staff and equipment, to the less obvious, such as the cost of cleaning buildings and clothes more frequently and the cost of replacing or repairing equipment or parts of buildings or other structures that have been severely corroded as a result of the acidic properties of some pollutants. Considerable expenditure is also incurred by industrialists and the government in minimising air pollution, sometimes because of the inappropriate siting of industrial or residential buildings".*

A study carried out for the American Lung Association concluded that national health costs of between US$ 443,000 million and US$ 9,349,000 million a year resulting from automotive and lorry exhaust pollution could be avoided (American Lung Association, 1990).

* *This chapter was prepared by D. Mage and M.P. Walsh*

Table 6.1 Percentages of motor vehicles in Bangkok exceeding a noise standard of 100 dB(A) at different locations

Site	Motor cycles	Diesel cars	Mini-buses	Buses	Lorries	Motor tricycles
A	21.4	5.3		81.0		
B	5.9	33.3		81.8	68.1	
C	6.0	20.0	78.6	90.7	86.8	70.0
D	3.1	5.6	33.3	88.6	60.0	13.3
E	10.2		41.4	82.4	88.0	
F	5.4	30.8		84.8	72.5	
G						9.4
H	6.7	20.0	40.5	77.8	62.7	
I	7.5	16.7	27.6	80.9	77.4	
J	5.1		63.3			

Source: National Environment Board of Thailand, 1990

The traffic noise pollution problem in most Asian cities is also severe, exceeding guidelines for hearing protection. For example, the noise pollution problem in Bangkok is caused mainly by heavy traffic. Alongside densely travelled roads equivalent sound pressure levels L_{eq} (see Chapter 3) for 24 hours have been found to be 75–80 dB(A), which is much greater than the US EPA recommended 70dB(A) for long-term hearing protection (National Environment Board of Thailand, 1990). The percentages of motor vehicles in Bangkok exceeding a noise standard of 100 dB(A) at a distance of 0.5 m is summarised in Table 6.1. Significant numbers of all vehicle categories tend to be very noisy but the problem is especially widespread with lorries and buses.

In this chapter, examples of the urban problems caused by motor vehicle proliferation are presented, with an emphasis on Asia where the growth in vehicle numbers is greatest. Several case studies are developed in order to define the problems and to give examples of various approaches to their solutions. These case studies are presented from Asia (Bangkok, Taipei and Manila) and from North and Central America (Mexico DF and Los Angeles). A detailed case study from Europe (Geneva) is presented in Chapter 7. Although traffic in African cities, such as Lagos and Soweto, presents its own set of pollution problems (Bolade, 1993; Khosa, 1993), insufficient detailed information exists to include a case study for Africa in this chapter.

Vehicular-related air pollution in the megacities of Asia (Bangkok, Beijing, Bombay, Calcutta, Delhi, Jakarta, Karachi, Seoul, Shanghai, Taipei and Tokyo) has been shown to exceed WHO guidelines and to present a serious public health problem (WHO/UNEP, 1992).

Chapter 5 summarised the status of technologies and other strategies currently available to reduce vehicle emissions. In spite of the technological advances which are now readily available, several countries in Asia have only been able to make limited progress in reducing vehicle emissions. Nevertheless, over the last five years a great deal of progress has been made in Asia and specific examples illustrating the efforts to combat emissions are summarised below.

6.1 Bangkok, Thailand

According to Stickland (1993) traffic conditions in Bangkok are generally recognised as being amongst the worst in the world. For example, traffic volumes on the main roads have been growing recently at 15–20 per cent per year, representing a doubling in volume every four to five years. More than 500 new vehicles come into use in Bangkok daily, and average traffic speeds of 10 km h^{-1} drop to a mere 5–6 km h^{-1} in peak traffic periods (Stickland, 1993).

In Thailand, the Office of the National Environment Board has monitored levels of CO, SPM and Pb near major roads in Bangkok since 1984. According to their 1990 annual report air pollution problems in Thailand are serious near major streets (National Environment Board of Thailand, 1990). In certain areas of the city where traffic is heavy, particulate matter concentrations far exceed the daily ambient air quality standard of 330 μg m^{-3} on any day, and are as high as two to three times standard values on some days. Carbon monoxide levels were also high, exceeding air quality standards in some congested areas. A study of blood Pb in policemen at three different rates of exposure to vehicular traffic, found a statistically significant positive link between traffic exposure and blood Pb concentrations (Raungchat, undated). Although reported blood Pb concentrations in Bangkok vary between 16 and 40 μg Pb per 100 ml, even the lowest reported average is three times as great as averages found it the USA and Western Europe (TDRI, 1990). Other air pollutants such as NO_2 and O_3 from photochemical oxidant reactions were found to at low levels in Bangkok due to the favourable meteorological conditions, with prevailing seasonal monsoon winds and a sea breeze. A breakdown of emissions for Bangkok is summarised in Figure 6.1. Motorcycles are a major contributor to HC and organic SPM levels and a significant source of CO and Pb. Diesel vehicles are the major source of sulphate and a significant source of carbonaceous SPM. Different vehicle categories, therefore, must be targeted to address different aspects of the

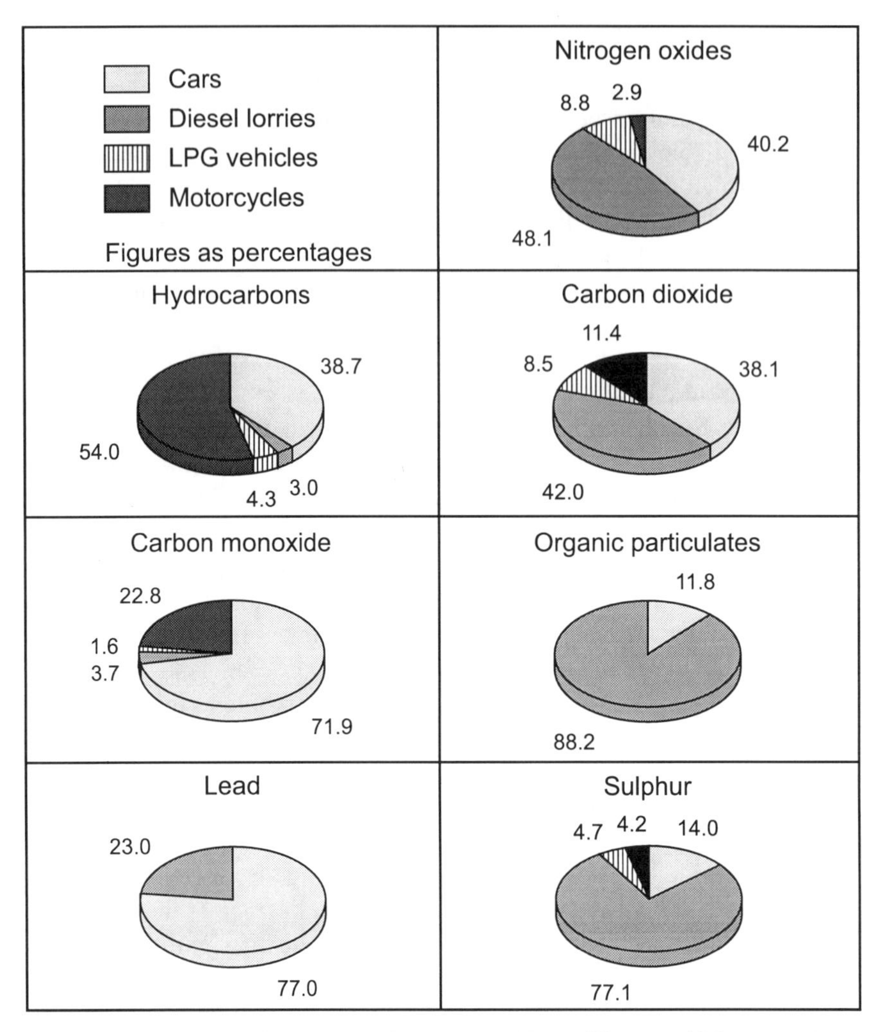

Figure 6.1 Proportion of major emissions attributable to different vehicle categories in Bangkok, 1989 (Data provided by National Environment Board of Thailand, 1990)

overall vehicle pollution problem. Figure 6.2 shows the likely impact of various control strategies on reducing emissions in Bangkok. The first priority is to restrain growth in future vehicle numbers. However, even if the overall growth could be constrained to only 5 per cent per year, vehicle emissions in Bangkok would increase dramatically by the year 2005, as illustrated in Figure 6.2. Consequently, a series of additional strategies is necessary in addition to restraint in the growth in vehicle numbers.

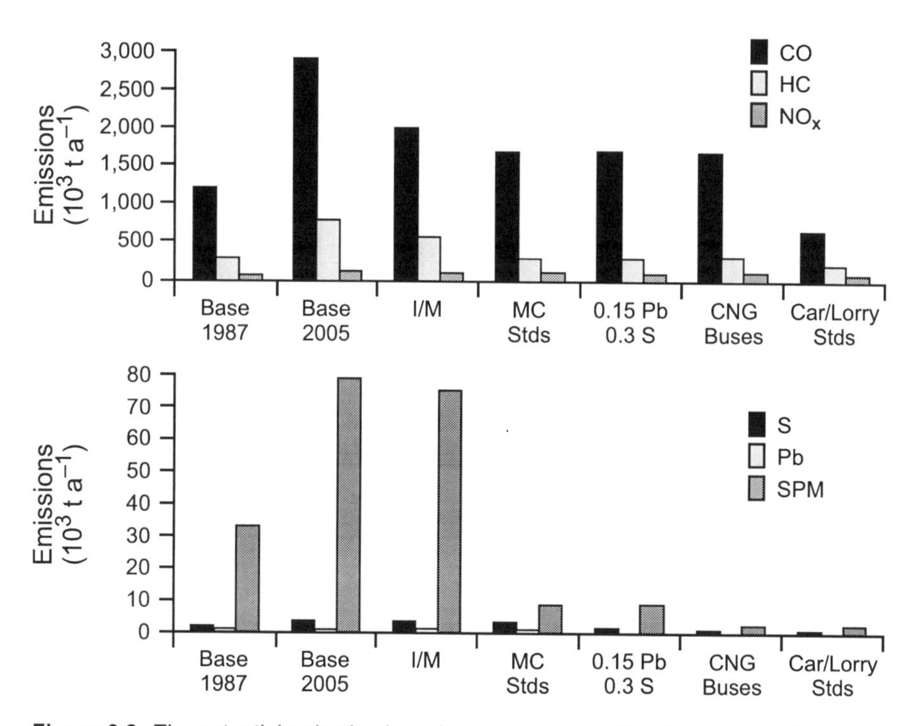

Figure 6.2 The potential reduction in major emissions resulting from introducing various vehicle pollution control strategies in Bangkok (Data provided by the National Environment Board of Thailand, 1990)

Based on a review of available air quality data, it has been estimated that roadside emissions of SPM, CO and Pb must be reduced by 85 per cent, 47 per cent and 13 per cent respectively, if acceptable air quality is to be achieved in Bangkok. Recent data indicate that O_3 concentrations downwind of the city may also be approaching unhealthy levels. It would be appropriate, therefore, to adopt measures which will reduce HC and NO_x emissions, as well as the O_3 precursors. In response to a serious air pollution threat, Thailand's current Seventh Plan has placed a high priority on improving air quality. Definite targets have been set to control the amount of SPM, CO and Pb present in Bangkok's major streets.

6.1.1 Current programme

A number of measures have been adopted to mitigate air pollution problems, particularly those caused by the transport sector. These measures are aimed not only at exhaust gas emission controls but also at the improvement of fuel and engine specifications, at the implementation of in-use vehicle I/M programmes, at public transport improvement through mass transit systems, and the improvement of traffic conditions through better traffic management. Measures directed towards reducing vehicle emissions include:

- Introduction of unleaded petrol at prices below that of leaded petrol (introduced in May 1991).
- Reduction of the maximum allowable Pb in petrol from 0.4 to 0.15 g l^{-1} (effective as of 1 January 1992).
- A decision to phase out leaded petrol by 1 January 1996.
- Reduction of the sulphur content of diesel fuel from 1.0 to 0.5 per cent (weight) as of April 1992 in the Bangkok Metropolitan Area and after September 1992 throughout the whole country. The use of low sulphur diesel fuel has been mandatory in Bangkok since September 1993.
- Reduction of the 90 per cent distillation temperature of diesel fuel from 370 °C to 357 °C as of April 1992 in the Bangkok Metropolitan Area and after September 1992 throughout the whole country.
- A requirement for all new cars with engines larger than 1,600 cc to meet the ECE R-83 standards after January 1993. All cars were required to comply after 1 September, 1993.
- Taxis and tuk-tuks have already been largely converted to operate on liquefied petroleum gas (LPG).
- ECE R40 requirements for motorcycles were introduced in August 1993 and were followed soon afterwards by ECE R40.01. The government decided on a third control step which was phased-in in 1995.
- ECE R49.01 standards for heavy duty diesel engine vehicles were put into effect.
- The government decided to reduce the sulphur level in diesel fuel from the previous 0.5 per cent (weight) to 0.25 per cent by 1996 and 0.05 per cent by the year 2000.

Currently, noise and emission testing are required and are conducted under the Land Transport Department's (LTD) general vehicle inspection programme. All new vehicles are subject to such inspection. For in-use vehicles, only those registered under the Land Transport Act (buses and heavy duty lorries) and commercial vehicles registered under the Motor Vehicles Act (taxis, tuk-tuks and rental vehicles) are subject to inspection during annual registration renewals.

Since 1996 the LTD has required all in-use vehicles to be inspected. Vehicles in use for 10 or more years are subjected to an annual inspection, while newer vehicles are subjected to inspection at different time periods, determined by the LTD. Private inspection centres are being licensed.

6.1.2 Future plans

In October 1995, Thailand began using 0.25 per cent sulphur in diesel fuel. Thailand's refineries, including Thai Oil, could produce 0.05 per cent sulphur in the diesel if they used more costly, low-sulphur crude oils. Another oil company in Thailand, Rayond and Star, could also produce low sulphur diesel fuel but do not have appropriate storage facilities at the present time. In an effort to help curb air pollution, the government of Thailand is planning to begin importation of 25,000–30,000 t of 0.05 per cent sulphur grade diesel every three months. Market enquiries were undertaken by the State owned petroleum authority to begin importation by 1997 of the low sulphur diesel for public buses throughout the country. While importation is, at the present time, the only solution, regional suppliers are limited. Exports from Japan have only been available since August 1997, while South Korea will not have enough to export until 1998. Singapore has a refinery capable of producing 0.05 per cent sulphur diesel oil, but does not have enough storage capacity.

6.1.3 Conclusions

Bangkok, like many other megacities in the world, has serious problems associated with the use of energy in the transport sector. Several factors, including population growth and rapid economic expansion are fundamental factors requiring consideration for long-term planning. Rapid industrialisation and urbanisation, coupled with the lack of land-use planning in the past, has contributed to the atmospheric pollution associated with the transport sector. This problem has been intensified by road infrastructures incapable of absorbing the rapidly growing vehicle population which, in turn, causes congestion. The lack of a mass transport system offering a good substitute for private vehicles encourages people to rely more on private vehicles, which has further contributed to the congestion problem.

It has been recognised that these problems can be alleviated by several measures, including: source reduction through improvement of fuel quality, I/M programmes, new vehicle standards, and traffic and demand management (such as having a good mass rapid transit (MRT) system). A great deal of work remains to be done, especially in respect of policy to control travel demand (i.e. demand side management).

6.2 Singapore

In Singapore, motor vehicle emissions are a significant source of air pollution. Vehicle numbers have been steadily increasing over the past decade as a consequence of rapid urbanisation and economic growth. At the beginning of 1993 there were approximately 550,000 vehicles.

6.2.1 Land transport policy

Singapore's land transport policy strives to provide free-flowing traffic within the constraint of a limited land area. A four-pronged approach has been adopted to achieve this. Firstly, the need to travel is minimised through systematic town planning. Secondly, an extensive and comprehensive network of roads and expressways, augmented by traffic management measures, has been built to provide quick accessibility to all parts of Singapore. Thirdly, a viable and efficient public transport system that integrates the MRT and bus services, is being promoted. Finally, the growth in, and use of, vehicles are being managed to prevent congestion on the road.

6.2.2 Mobile source controls

Singapore's strategy for reducing pollution from motor vehicles has two main aspects: improving engine and fuel quality to reduce emissions and using traffic management measures to control growth in vehicle numbers and fuel consumption. The Pollution Control Department works closely with the Registry of Vehicles to implement this strategy.

The permissible sulphur content in automotive diesel was reduced from 0.5 per cent to 0.3 per cent by weight with effect from 1 July 1996. There are also plans to reduce the sulphur content to 0.05 per cent by weight in the longer term. This would further reduce the emissions of particulate matter and sulphur dioxide from diesel vehicles and also prepare the way for the introduction of more stringent emission standards for diesel vehicles which would include a requirement for the use of catalytic converters. The emission standards for petrol vehicles have been progressively tightened since 1984 and the standards in force since 1 July 1994 are the Consolidated Emissions Directive 91/441 and the Japanese emission standards (Article 31 of Safety Regulations for Road Vehicles). Since October 1991, motorcycles and scooters have been required to comply with the emission standards stipulated in the US Code of Federal Regulation 86.410–80 before they can be registered for use in Singapore. Since 1 July 1997 emission standards applicable to passenger cars and light commercial vehicles follow the European Union Directive 93/59/EEC and up to 30 June 1998 the Japanese standards JIS 94 (the Japanese standards were accepted for only one year to allow motor traders more time to comply with the EU standards).

Singapore has, between 1980 and 1987, gradually reduced the Pb content of leaded petrol from 0.84 $g\ l^{-1}$ to the current level of 0.15 $g\ l^{-1}$. Unleaded petrol was introduced in 1991 and its use is encouraged by a differential tax system making unleaded petrol about 10 Singapore cents per litre cheaper than leaded petrol. At the end of 1997, the sale of unleaded petrol constituted abut 75 per cent of the total petrol sales. Availability of unleaded petrol has enabled Singapore to adopt more stringent exhaust emission standards for petrol-driven vehicles which require the use of catalytic converters. The oil companies have voluntarily agreed to phase out leaded petrol by July 1998.

The current in-use vehicle emission standards are:

- For petrol vehicles with registration dates of before 1 October 1986, on or after 1 October 1986, and on or after 1 July 1992, at engine idling CO emissions should be 6 per cent, 4.5 per cent and 3.5 per cent by volume respectively.
- For diesel vehicles, smoke emissions should be below 50 Hartridge Smoke Units (HSU) at free acceleration.

The frequency of vehicle inspection is yearly for cars and motorcycles older than three years and for goods vehicles less than 10 years old. Taxis less that 10 years old must be inspected every six months, as must buses of any registration date and good vehicles less than 10 years old.

Traffic management measures

The situation in Singapore is unique because it is essentially a city state with a large population living on a small land mass. Urbanisation, industrialisation and infrastructure development are still progressing, fuelled by a growing economy. With such a combination of factors, there is potential for serious environmental problems from stationary and mobile sources if these sources are not managed or controlled properly. In the case of motor vehicles, the need to control their impact on traffic flow and the environment has given rise to a unique set of traffic management measures.

Vehicle registration and licensing. The expense of owning and operating a vehicle in Singapore has effectively slowed the growth in vehicle numbers. Car owners wishing to register their cars must pay a 45 per cent import duty on the car's open market value (OMV), a registration fee of S$ 1,000 for a private car (S$ 5,000 for a company-registered car) and an Additional Registration Fee (ARF) of 150 per cent of the OMV. In addition, car owners pay annual road taxes based on the engine capacity of their vehicles. The road tax of company-registered cars is twice as high as for individually owned cars. For diesel vehicles, a diesel tax which is six times the road tax of an equivalent petrol vehicle is payable.

To encourage people to replace their old cars with newer, more efficient models, a Preferential Additional Registration Fee (PARF) system was introduced in 1975. Private car owners who replaced their cars within 10 years were given PARF benefits that they could use to offset the registration fees they had to pay for their new cars. For cars registered on or after 1 November 1990, the PARF benefits varied according to the age of the vehicle at deregistration. For cars registered before 1 November 1990, a fixed PARF benefit was given upon deregistration, based on the engine capacity of the car. To provide a higher PARF benefit to car owners who deregistered their cars before 10 years, all PARF-eligible cars registered on or after 1 November 1990 received higher fees if the vehicle was newer.

Vehicle quota system. High taxes alone would not have ensured that vehicle numbers grew at an acceptable rate, therefore a vehicle quota system was introduced. After 1 May 1990, any person who wished to register a vehicle first had to obtain a vehicle entitlement in the appropriate vehicle class, through a bidding system. Tendering for specified numbers of vehicle entitlements was conducted monthly. Successful bidders paid the lowest successful bid price of the respective category in which they had bid. A vehicle entitlement is valid for 10 years from the date of registration of the vehicle.

If the owner wished to continue using the vehicle on expiration of the vehicle entitlement, it was necessary to revalidate the entitlement for another 5 or 10 years, by paying a revalidation fee (pegged at 50 per cent or 100 per cent of the prevailing quota premium respectively).

Weekend car scheme. The weekend car scheme was introduced on 1 May 1991 to allow more people to own private cars without adding to traffic congestion during peak hours. Cars registered under the scheme enjoyed substantial tax concessions, including a 70 per cent reduction in road tax and a tax rebate up to a maximum of S$ 15,000 on registration. Weekend cars were identifiable by their red license plates, fixed in place with a tamper-evident seal. They could only be driven between 1900h and 0700h during the week, after 1500h on Saturdays and all day on Sundays and public holidays.

Weekend cars can also be driven outside those hours but the owners were required to display a special day license. Each weekend car owner is given five free day licenses per year and could buy additional ones at S$ 20 each.

Area licensing scheme. The Area Licensing Scheme (ALS) was introduced in June 1975 to reduce traffic congestion in the city area during peak hours. Originally only passenger cars were affected, but the scheme has gradually been modified to include all vehicles except ambulances, fire engines, police vehicles and public buses.

Public transportation. Public transport in Singapore is widely available and includes an MRT system, a comprehensive bus network and over 13,000 taxis.

6.2.4 Conclusions

Besides technical control measures (controls on engines and fuel quality), the use of traffic control measures has contributed significantly to the protection of air quality in Singapore. Although the present measures appear to be adequate, Singapore will continue to look ahead for ways of improving them further. Pilot studies of three electronic road pricing systems have been carried out in Singapore and the most suitable system was due to be selected for implementation in 1997.

6.3 Hong Kong

In Hong Kong, the EPA reported that in 1989 approximately 1.5–2 million people were exposed to unacceptable levels of SO_2 and NO_2 and about 3 million people were exposed to high SPM levels. Many people were exposed to unacceptable levels of all three pollutants (Hong Kong EPA, 1989). The high levels of air pollution in many parts of Hong Kong are seriously aggravating the health of those already suffering from respiratory illnesses and contribute to the onset of chronic health problems.

Hong Kong's vehicle pollution control efforts continue to focus on diesel SPM control because SPM is the most serious pollution problem at present. Motor vehicles are estimated to be responsible for approximately 50 per cent of the PM_{10} emissions.

6.3.1 Current programme

As of 1 April 1997, the sulphur level in diesel fuel was reduced to 0.05 per cent (weight). Diesel vehicle emission standards were also tightened by this date. Since 1 April 1997, all new passenger cars and taxis have had to comply with either the USA 1988 standards (SPM = 0.12 g km^{-1}, NO_x = 0.63 g km^{-1}) or the EU Step 1 standards (93/59/EEC: SPM = 0.14 g km^{-1}, HC and NO_x = 0.97 g km^{-1}) or the Japanese 1994 standards (SPM = 0.34 g km^{-1}, NO_x = 0.72 g km^{-1} for vehicles weighing less than 1.265 t or 0.84 g km^{-1} for those above this weight). As of 1 October 1998, new private cars should comply with Californian 1994 standards or, alternatively, from 1 October 2000 onwards the EU Step 3 emission standards must be applied. The emissions of taxis should comply with EU Step 2, or USA 1996, or Japanese 1998 standards from 1 April 1999 (Walsh, 1997); alternatively, from 1 October 2000 onwards, the emission standards of EU Step 3 will also be applied. Similar requirements will apply to all light and medium goods vehicles and light buses. For goods vehicles and buses with a design weight of 3.5 t or more, either the 1991 USA (SPM = 0.80 g/kWh, NO_X = 8.04 g/kWh) or the EU Step 1 standards (SPM = 0.61 g/kWh for engines producing less than 85 kW or 0.36 g/kWh for engines producing more; NO_X = 8.0 g/kWh for all engines) have applied since 1 April 1995. For goods vehicles produced after 1 April 1997 the more stringent standards of EU Step 2 or the 1994 USA standards apply. It is proposed to replace the latter standards for new goods vehicles in this weight class by the 1998 USA standards from 1 October 1998 onwards; it is further proposed to apply the EU Step 3 or the 1998 USA emission standards from 1 October 2000 onwards.

In-use smoke limits based on the EU free acceleration test (72/306/EEC) will be lowered to 50 HSU; in certification, the limits will be 40 HSU. As a result of a HK$ 1 per litre price reduction for unleaded petrol compared with leaded, unleaded petrol is now responsible for 71 per cent of total petrol sales. The benzene content of the unleaded petrol is only 3.44 per cent, which is virtually the same as leaded petrol.

Since 1 April 1997, the maximum Pb content of petrol has been lowered to 0.005 per cent, with a sulphur content of less than 0.05 per cent. Either EU Step 2 or USA 1994 or Japan 1978 exhaust emission standards have also had to be applied to new petrol driven cars since that date. It is proposed to strengthen these standards to EU Step 3 or USA 1996 standards by 1 October 2000. The Japan 1978 standards will be tightened to include evaporative emission standards according to EU, USA or Japanese regulations. Similar requirements are applicable or are being proposed for goods vehicles. Extensive summary tables on exhaust emission regulations in Honk Kong for petrol and diesel-driven vehicles can be found in Walsh (1998).

6.3.2 Future plans

An analysis of the motor vehicle-related urban SPM problem indicates that 17 per cent comes from buses, 63 per cent from goods vehicles and the remainder from all vehicles under 5.5 t. As a matter of policy, Hong Kong is still trying to convert all light duty diesel vehicles, including taxis, to using petrol. Analyses are also being carried out regarding the possibility of converting some or all taxicabs to either CNG or electric power. Hong Kong also remains interested in the possibility of retrofitting buses with either catalysts or diesel SPM filters. A proposal has been submitted to the Asia–USA partnership to fund such an effort and discussions have been initiated with potential suppliers in Europe.

Table 6.2 Emission standards for new petrol and LPG vehicles in the Republic of Korea, 1987–2000

Vehicle type	Date of implement-ation	Test	CO ($g\ km^{-1}$)	NO_x ($g\ km^{-1}$)	Exhaust HC ($g\ km^{-1}$)	Evapor-ative HC (g per test)
Small size	1 July 1987	CVS-75	8.0	1.50	2.10	4.0
cars (< 800 cc)	1 July 2000	CVS-75	2.11	0.62	0.25	2.0
Passenger	1 Jan. 1980	10 mode	26.0	3.00	3.80	
cars	1 July 1984	10 mode	18.0	2.50	2.80	
	1 July 1987	CVS-75	2.11	0.62	0.25	2.0
	1 Jan. 2000	CVS-75	2.11	0.25	0.16	2.0
Light duty						
lorries (< 3 t)	1 July 1987	CVS-75	6.21	1.43	0.50	2.0
(< 2 t)	1 Jan. 2000	CVS-75	2.11	0.62	0.25	2.0
(2 t – 3 t)	1 Jan. 2000	CVS-75	6.21	1.43	0.50	2.0
Heavy duty	1 Jan. 1980	6 mode		1.6%	2,200 ppm	520 ppm
vehicles	1 July 1987	US transient	15.5	10.7	1.3	4.0
	1 Feb. 1991	13 mode	33.5	11.4	1.3	
	1 Feb. 2000	13 mode	33.5	5.5	1.4	

Source: Walsh, 1994

6.4 Republic of Korea

A series of recent amendments in the Air Quality Control Law will gradually tighten vehicle emissions standards in the Republic of Korea as summarised in Tables 6.2 to 6.4.

The sulphur level in diesel fuel was reduced to a maximum of 0.4 per cent (weight) during the period from 2 February 1991 to 31 December 1992 and to 0.2 per cent (weight) during the period from 1 January 1993 to 31 December 1995 and to 0.1 per cent (weight) thereafter.

The Republic of Korea is also investigating possible improvements in their I/M programme, including the possible addition of the new short test procedure (IM240, based on the first 240 seconds of the USA Federal Test Procedure) developed by the US EPA.

Research into the use of diesel SPM filters is ongoing. Three types of approaches are under investigation: burner systems which are seen as the prime choice for large vehicles; electrically heated systems which are seen as the prime choice for medium sized vehicles; and Cerium fuel additive systems which are seen as the prime choice for smaller vehicles. Research is also being performed in Korea on electrically heated catalysts, CNG engines, two stroke engines and lean NO_X catalysts.

Table 6.3 Emission standards for new diesel vehicles in the Republic of Korea

Vehicle type	Date of Implementation	Test	CO (ppm)	NO_x (ppm)	HC (ppm)	PM (ppm)	Smoke (%)
Passenger cars, light duty lorries (< 3 t), heavy duty vehicles	1 Jan. 1980	Full load					50
Direct injection							
Passenger cars	1 July 1984	6 mode	980	1,000	670		50
Light duty lorries (< 3 t)	1 Jan. 1988	6 mode	980	850	670		50
Heavy duty vehicles	1 Jan. 1993	6 mode	980	750	670		40
Indirect injection							
Passenger cars	1 July 1984	6 mode	980	590	670		50
Light duty lorries (< 3 t)	1 Jan. 1988	6 mode	980	450	670		50
Heavy duty vehicles	1 Jan. 1993	6 mode	980	350	670		40

Source: Walsh, 1994

Table 6.4 Emission standards for new diesel vehicles in the Republic of Korea

Vehicle type	Date of Implementation	Test	CO ($g\ km^{-1}$)	NO_x ($g\ km^{-1}$)	HC ($g\ km^{-1}$)	PM ($g\ km^{-1}$)	Smoke (%)
Passenger cars	1 Jan. 1996 1 Jan. 2000	CVS-75 CVS-75	2.11 2.11	0.62 0.62	0.25 0.25	0.08 0.05	
Light duty lorries (< 3 t)	1Jan. 1996	CVS-75	6.21	1.43	0.50	0.31	
Light duty lorries (< 2 t)	1 Jan. 1996 1 Jan. 2000	CVS-75 CVS-75	6.21 2.11	1.43 0.75	0.50 0.25	0.16 0.12	
All other light duty lorries	1 Jan. 2000	CVS-75	6.21	1.00	0.50	0.16	
Heavy duty vehicles	1 Jan. 1996 1 Jan. 2000	13 mode 13 mode	4.90 4.90	11.00 6.00	1.20 1.20	0.90 0.25	35 25
City buses	1 Jan. 2000	13 mode	4.90	6.00	1.20	0.10	25

Source: Walsh, 1994

The diesel retrofit project is proceeding smoothly in Seoul. At the end of 1996, the Seoul Metropolitan Government bought 1,481 trap systems (catalysts, electric heaters, burners) for rubbish collection vehicles. These systems were installed by July 1997. The Korean EPA was planning to install approximately 20,000 trap systems for city buses, rubbish collection vehicles and government-owned buses and lorries in 1997. All petrol and LPG fuelled vehicles have had to be tested for CO and HC, as well as for the A:F ratio, since the beginning of 1997. Loaded emission testing is being considered for application in a few years time.

6.5 Taipei, Taiwan

As a result of rapid industrialisation and motorisation, Taipei is experiencing severe environmental problems. The worst air pollutant is SPM. Ozone is also a serious problem and CO concentrations are excessive. Vehicles are the dominant source of emissions in Taipei, with NO_X, HC and CO emitted from motor vehicles accounting for about 50 per cent of all emissions (Shen and Huang, 1989). An especially important source, fairly unique to Taipei, is motorcycles. Taiwan has the highest density of motorcycles in the world and, because 80 per cent of motorcycles are equipped with two-stroke engines, they contribute about 33 per cent and 10 per cent of HC and CO emissions respectively. To improve the air quality significantly in the Taiwan area, vehicular pollution, especially from motorcycles, must be controlled (Shen and Huang, 1989).

In their analysis of the air quality problem, the Taiwan EPA has developed a comprehensive approach to motor vehicle pollution control. Building on its early adoption of US EPA 1983 standards for light duty vehicles (starting 1 July 1990) it moved to US EPA 1987 requirements, (which include the 0.2 gram per mile SPM standard), as of 1 July 1995. Heavy duty diesel SPM standards almost as stringent as US EPA 1990 standards (6.0 g/bhph NO_X and 0.7 g/bhph SPM, using the transient test procedure of the USA), went into effect on 1 July 1993. It was intended that US EPA 1994 standards (5.0 g/bhph NO_X and 0.25 g/bhph SPM) would be adopted, probably for introduction by 1 July 1997.

Diesel fuel contained 0.3 per cent of sulphur in the early 1990s. A proposal to reduce sulphur levels to 0.05 per cent by 1997 was considered. The Executive Yuan approved, on 10 December 1992, increases of up to 1,700 per cent for the amount of fines to be levied against motorists who violated the Air Pollution Control Act. The new fine took effect in early 1993 following official public notice and raised the maximum fine for motor vehicle pollution from US$ 138 to US$ 2,357. All forms of motorised transportation were included in the new fine schedule, including aeroplanes, boats, and power water skis.

The most distinctive feature of the Taiwan programme was the effort to control motorcycle emissions, reflecting the fact that motorcycles were the dominant vehicle and were a substantial source of emissions. The first standards for new motorcycles were imposed in 1984 at 8.8 g km^{-1} for CO and 6.5 g km^{-1} for HC and NO_X combined, using the ECE R40 test procedure (Shen and Huang, 1989). In 1991, the limits were reduced to 4.5 g km^{-1} for CO, and 3.0 g km^{-1} for HC and NO_X combined. These requirements were phased in over two years and by 1 July 1993 were applied to all new motorcycles sold in Taiwan. As a result of these requirements, the engines of four-stroke motorcycles have been redesigned to use secondary air injection. All new two-stroke motorcycles were fitted with catalytic converters. Since 1992, electric motorcycles have been available but sales have been modest.

Motorcycle durability requirements have been imposed since 1991. All new motorcycles tested since then have been required to demonstrate that they can meet emissions standards for a minimum of 6,000 km. All new motorcycles must also be equipped with evaporative controls.

In order to reduce pollution from in-use motorcycles, the Taiwan EPA has been actively promoting a motorcycle I/M system. In the first phase, from February to May 1993, the Taiwan EPA tested approximately 113,000 motorcycles in Taipei City. Of these, 49 per cent were given a blue card indicating that they were clean, 21 per cent a yellow card indicating that their emissions were marginal, and 30 per cent failed. Between December 1993 and May 1994, approximately 142,000 motorcycles were inspected with 55 per cent receiving blue cards, an increase of 6 per cent from the earlier programme, and 27 per cent failed, a decrease of 3 per cent. The major remedial action for motorcycles that failed was the replacement of the air filter, at an average cost of US$ 20.

In continuing its regulations for the control of motorcycle emissions, the Taiwan EPA has adopted the Third Stage Emission Regulation which is to be implemented from 1998. The new standards will lower CO to 3.5 g km^{-1} and HC and NO_X to 2 g km^{-1}. In addition, the durability requirement will be increased to 20,000 km. Finally, the market share for electric powered motorcycles will be mandated at 5 per cent. The EPA will also extend the periodic motorcycle I/M programme.

6.6 Manila, the Philippines

Private motor vehicles and motor cycles in Metropolitan Manila are primarily petrol driven whereas taxis, jeepneys and buses providing public transportation are mostly fuelled by diesel. Of the 510,000 vehicle registrations in 1988 for petrol and diesel vehicles, approximately 383,000 (or 75 per cent) were petrol fuelled. The number of motor vehicles is expanding at a far greater rate than the road system can cope with. In 1993–94, when the number of newly registered cars increased by 53 per cent, utility vehicles by 44 per cent and

Table 6.5 Results of a cohort study on pollutant variables in Metropolitan Manila

Pollutant variable	Cohort: Jeepney driver's children	School children	Street vendors	WHO guidelines
SPM (μg m^{-3}) 24 h average	550	314	345	120
Pb blood (μg/100 ml)	19.9	14.0	17.7	10
COHb (%)	22			2

Source: Torres and Subida, 1994

lorries by 29 per cent, the length of roads increased by only 0.4 per cent. Exposures to automotive pollutants had never been measured until recent years, when Torres and Subida (1994) performed an epidemiological study of illness among jeepney drivers, air conditioned bus drivers, school children, child street vendors (7–14 years of age) and commuters. The study reported mean SPM exposure (obtained by personal monitoring), blood Pb for the children and jeepney drivers, and blood COHb for the jeepney drivers (Table 6.5).

The values given in Table 6.5 are well above the WHO guidelines and Philippines national standards for these pollutant variables. They are indicative of the severity of the automotive exposures of people who have occupational contact with automotive exhaust and which can lead to chronic health problems. The high COHb found in jeepney drivers, many of whom were chain smokers, represents a 7 per cent increment above the expected value of 15 per cent COHb found in other adult chain smokers, which in turn is 13 per cent higher than the 2–3 per cent COHb in non-smokers from the Manila area. The mean values of blood Pb concentrations in the children were well above the USA Centre for Disease Control (CDC) recommended limit of 10 μg per 100 ml, which was exceeded by more than 80 per cent by both cohorts of children. Comparison of pulmonary function between Metro Manila and rural school children showed that the urban children had higher blood Pb levels than the rural children (decrements from the expected values for sex, height and age) by a factor of two.

6.7 Mexico City, Mexico

The number of automobiles in Mexico City has grown dramatically in the past few decades; there were approximately 48,000 cars in 1940, 680,000 in 1970, 1.1 million in 1975 and 3 million in 1985. The Mexican government estimated, that in 1990, fewer than half of these cars were fitted with even modest pollution control devices. Virtually none were equipped with state-of-the-art exhaust after-treatment systems at that time. In addition, more than 40 per cent of the cars were over 12 years old and, of these, most had engines in need of major repairs. The degree to which the existing vehicles are in need of maintenance is reflected by the results of the "voluntary" I/M programme run by the City of Mexico during 1986–88. More than 600,000 vehicles were tested (209,638 in 1986, 313,720 in 1987 and 80,405 in the first four months of 1988) and of these about 70 per cent failed the petrol vehicle standards and 85 per cent failed the diesel standards (65 HSU) (Mage and Zali, 1992).

Although there were not as many buses as cars, buses were considered a major pollution source because of the high levels of fine particles produced by their diesel engines and because of the public exposure to these emissions. On a per capita basis in 1991, there was only one car for every 16 people in Mexico, whereas in Brazil, for example, there was only one car for 13 people (compared with one car for every 1–2 people in industrialised countries). This illustrates that there is still a tremendous potential for increased in vehicle numbers in Mexico, above and beyond the projected growth in the overall population.

Not surprisingly, with such a large and growing vehicle population, and with such limited pollution control for these vehicles, Mexico City is classed amongst the most polluted cities in the world. The problem is compounded by stagnant meteorological conditions throughout the winter season, by its high elevation, and by its physical location (i.e. in a geographical bowl surrounded by mountains).

The air pollution problem in Mexico City is probably aggravated, from a public health standpoint by the commuting patterns which exist there. More than 20 per cent of workers spend three or more hours commuting each day and for 10 per cent daily travel time exceeds five hours. Some employees live as far as 80 km from their place of work (Fernandez-Bremauntz, 1992; Fernandez-Bremauntz and Merritt, 1992). As shown in Chapter 4, commuters are among the proportion of the population most highly exposed to motor vehicle emissions. An increasingly intensive effort has been made to develop a comprehensive package of pollution control measures. Motor vehicle controls, including more stringent new car standards, retrofitting of some older vehicles and I/M, are important components of the package.

6.7.1 New vehicles

Mexican authorities introduced more stringent standards for light duty vehicles, culminating with full USA-equivalent automobile requirements by

Table 6.6 Emission standards for Mexico

	Emission standard (g/mile)					
	1989	1990	1991	1992	1993	1994
Cars, no lorries						
HC	3.20					
CO	35.20					
NO_x	3.68					
Private cars						
HC		2.88	1.12	1.12	0.40	0.40
CO		28.80	11.20	11.20	3.40	3.40
NO_x		3.20	2.24	2.24	1.00	1.00
Commercial vehicles (< 2.726 t)						
HC		3.20	3.20	3.20	3.20	1.00
CO		35.20	35.20	35.20	35.20	14.00
NO_x		3.68	3.68	3.68	3.68	2.30
Light duty lorries (2.727–3 t)						
HC		4.80	4.80	3.20	3.20	1.00
CO		56.00	56.00	35.20	35.20	14.00
NO_x		5.60	5.60	3.68	3.68	2.30

Source: Brosthaus *et al.*, 1994

Table 6.7 Heavy duty vehicle emissions standards in Mexico

Model year	HC (g/bhph)	CO (g/bhph)	NOx (g/bhph)	PM (g/bhph)
1993	1.3	15.5	5.0	0.25
1994–97				
Large buses	1.3	15.5	5.0	0.07
Medium buses	1.3	15.5	5.0	0.10
1998 and later				
Large buses	1.3	15.5	4.0	0.05
Medium buses	1.3	15.5	4.0	0.10

Source: Brosthaus *et al.*, 1994

1993 (Table 6.6). Interim standards for 1989 through to 1992 were consistent with a proposal made by the automobile manufacturers. When these standards went into full effect, Mexico became the first Latin American country to introduce cars meeting USA standards. Starting in 1995, all cars, light commercial vehicles and light lorries were also required to meet an evaporative emission standard of 2.0 g per test. Heavy duty diesel truck and bus emission standards have also been adopted and are summarised in Table 6.7.

The new vehicle standard setting process proved to be adequate, but it had several problems. There were several categories of vehicles for which no standards applied. These included petrol fuelled heavy duty vehicles or engines, and motorcycles.

The standards adopted for new vehicles were less stringent, for virtually all vehicle categories, than those in effect in the USA. This difference will increase in the future unless standards are gradually tightened to keep pace with technological advances. Mexico is not getting the full benefits of technological advances, such as within-vehicle diagnostics, which could play an important role in minimising emissions from vehicles in use.

In 1996, the Government did not have a laboratory capable of carrying out emission tests on prototype or new vehicles, or for vehicles in use, to assure compliance with manufacturer directed standards. Certification was generally carried out by the vehicle manufacturers with some oversight by Government officials. Occasional certification tests were carried out at the Instituto Mexicano del Petroleo (IMP) laboratory, Mexico DF, at the request of either government or industry. In addition, some tests of vehicles, as they came off the assembly lines, were occasionally carried out at manufacturers' laboratories. This programme provided some assurance that vehicles were designed properly, but less assurance that they were built properly, and virtually no assurance that they were durable in use.

One possible remedy for the problems outlined above is the construction of a comprehensive emission testing laboratory. This is under active consideration in the State of Mexico but, while no such laboratory exists, the benefits of new vehicle standards are highly uncertain.

6.7.2 In-use vehicle retrofit

Mexico has a high percentage of old vehicles (as is typical of many developing countries) and therefore it will take many years before new car standards can have a very significant effect on the environment. Any short-term improvements are dependent on reducing emissions from the existing fleet. Based on the analysis by the Under-Secretariat of Ecology (SEDUE), strict application of a requirement to retrofit some of the vehicles (those capable of operating satisfactorily on unleaded petrol) would have the greatest pollution reduction potential during the 1990s. As a first step, SEDUE intends to require official cars dating since 1983 to be retrofitted. If this is successful, the next step will be to extend the requirement to all vehicles involved in transporting people, e.g. taxis and combis.

6.7.3 Inspection and maintenance

Up until the end of 1995 a dual I/M system had been in effect in Mexico City, combining both centralised (test only) facilities and decentralised private garages (combined test and repair). It became increasingly clear that the private garage system was not working. For example, it was observed that the failure rate in the private garages averaged about 9 per cent whereas in the centralised facilities, the failure rate was about 16 per cent. Stations conducting improper or fraudulent inspections were taken to court on several occasions, with the courts usually agreeing to close them down. By the end of 1995, as part of the development of the "Programa para Mejorar la Calidad del Aire en el Valle de Mexico, 1995–2000" (the New Programme), and in recognition of the critical role that I/M has and will play in the strategy for Mexico City, it was decided to close all the private garage inspection stations and to switch to a completely decentralized system. In addition, it was agreed in principle that the State of Mexico would proceed with the same approach but with a delayed schedule. The upgraded I/M programme has three main stages, and if fully implemented as planned, will be a substantial improvement over the original programme. Major improvements include:

- Inspections only at stations which do not have responsibility to repair vehicles (test only or centralised macro centres).
- Tighter standards for all vehicle categories.
- Improved, loaded mode testing of CO, HC and NO_X emissions.
- Independent auditors at each station to monitor the quality and integrity of each inspection.

Even these improvements are not adequate. Major problem areas which remain include:

- Evaporative emissions testing is not planned at present. Although it is scheduled to be added at some time in the future, no firm commitment or schedule has been made. Since evaporative emissions can be a substantial source of excess HC, this deficiency should be addressed.
- At some future time, but with no firm date, a similar transition is planned to take place in the State of Mexico. Because any vehicle can receive its inspection at any station in the Mexico City Metropolitan Area (MCMA), it is very likely that problem vehicles will tend to gravitate towards having their inspections in the place most likely to give the most lenient inspection, i.e. the SoM. This could lead to an immense loophole in the entire effort.
- Eventually, it is planned to set up a centralised computer system whereby each test at each facility is recorded in real time. This is not possible at present because of a lack of funds and available telephone lines.

- Another significant problem area is the lack of skilled mechanics. A very good mechanic training programme will be necessary to make this programme work.

6.7.4 In-use standards for exempting vehicles from the two day "hoy no circula" ban

The Government has made a commitment to announce the emission standards which must be met to qualify for an exemption from the two day "hoy no circula" ban during air pollution alerts. The likelihood was that these standards would be 2.0 per cent CO and 200 ppm HC. These standards have two objectives: to make sure that only clean vehicles are used during high pollution days so that pollution will be minimised, and to stimulate the modernisation of the fleet by encouraging people to replace their old, highly polluting vehicles with new ones which could qualify for the exemption.

As noted above, there was also an intention to replace the 1996 I/M test with a loaded test by 1 January 1997. Therefore, new standards for the exemption would need to be determined by that time, and should have been announced sufficiently in advance of the 1 January deadline in order to allow people to adjust their behaviour accordingly.

Standards to allow exemption from the normal "hoy no circula" programme were also expected. It is likely that values of 1.0 per cent CO and 100 ppm HC would be set. Adjustments were also due to be made later in the 1996 for the ASM test.

6.7.5 Fleet modernisation and alternative fuel conversions

Transport-related emission control measures in the Integrated Programme include replacement of existing taxis and combis with new vehicles meeting emissions standards and conversion of petrol-powered cargo lorries to LPG fuel, with the incorporation of catalytic converters.

A taxi renewal programme was launched by the Mexican authorities in 1991. In Mexico City, the aim of the programme was to replace over a two year period, all pre-1985 taxis with new taxis meeting current emissions standards. In the SoM, the initial aim was to replace all pre-1982 taxis with 1982 or newer vehicles. The medium-term aim was to facilitate ongoing renewal so that no MCMA taxi would be more than six years old. This would necessitate the replacement of approximately 55,000 vehicles (93 per cent of the fleet).

Within MCMA there were about 200,000 petrol and 50,00 diesel lorries, which together are the second largest mobile source of air pollution, (i.e. after cars). A regulation issued by the City of Mexico requires that all lorries built prior to model year 1977 must be replaced.

By the end of 1995, 25,000 vehicles had been converted to LPG using systems designed for low emissions. Almost all conversions included a three-way catalytic converter and closed loop control system. There are plans to continue the conversion of additional vehicles, although the rate of conversion has declined to only about 3,500 vehicles per year. Kits have been certified for over 200 models.

Mexico City was expecting to progress more quickly with the modernisation and conversion of its motor vehicles in the remainder of 1996. It planned to introduce 500 new, full-size buses as replacements for existing microbuses. A typical exchange would involve two microbuses exchanged for one full-sized bus. Of these buses, 10 per cent or 50 per cent were expected to use CNG. In addition, approximately 2,000 rubbish collection vehicles could be converted.

6.7.6 Fuels

Since 1994, fuels have improved significantly in Mexico City, according to PEMEX (Mexican National Petroleum Co.). The Pb content of leaded fuel has been reduced to about 0.15 g l^{-1}, sulphur concentrations have been reduced and a grade of unleaded petrol has been introduced. Furthermore, new detergents have been added to the petrol and are designed to reduce CO, HC and NO_X emissions.

Unfortunately, no independent government agency, such as SEDUE, has the authority or responsibility to ensure fuel quality independently. This allows and encourages fears that even proposed improvements in fuel quality, such as the elimination of Pb, may actually make environmental problems worse. Concerns have even been raised that the introduction of unleaded fuel has increased O_3 levels. Recently, it appears that PEMEX has raised Pb concentrations to the previous levels. Nevertheless, substantial progress in improving fuel quality has occurred to date and progress is continuing.

Sales of Magna Sin (unleaded petrol) are slowly increasing and in Mexico City this now represents approximately 46 per cent of total sales. In wealthier cities such as Monterey, it accounts for about 84 per cent of the total. In addition to its wealth, Monterey is closer to the border with the USA and thus attracts a greater proportion of cars assembled in the USA and which require unleaded petrol. The sulphur content of diesel fuel has been reduced from about 3 per cent to a maximum of 0.05 per cent by weight.

The price of leaded Nova petrol was 2.39 pesos per litre whereas for Magna Sin the cost was 2.47 pesos. In early May 1996, an additional surcharge was added to provide a second fund within the Trust Fund. The surcharge was 3 cents per litre for Nova and 1 cent per litre for Magna Sin, thereby lowering the differential between leaded and unleaded fuel to 6 cents per litre after the surcharge and the monthly price increase on petrol.

During the high pollution winter season, petrol specifications were lowered as follows:

	Previous specification	Winter specification
Olefins	15%	12%
Benzene	2%	1%
Aromatics	30%	25%
RVP	8.5 psi	7.8 psi

It was later decided, as part of the "New Programme" to retain the winter specifications throughout the summer; PEMEX agreed to this.

6.7.7 Improvements in ambient air quality

Air quality has improved significantly for a number of pollutants over recent years, although serious problems remain. Ozone and SPM are still the most serious problems.

Carbon monoxide

The highest CO level recorded in 1990 was 24 ppm (8-hour average), substantially exceeding the Mexican 8-hour average standard of 13 ppm, as well as the US EPA and WHO recommendations of 9 ppm for 8 hours. Human exposure studies have shown that pedestrians, roadside vendors, and vehicle drivers and passengers are exposed to much higher CO concentrations at street level. At the present time, even though the Mexican standard has been lowered to an 8-hour average of 11 ppm, very few if any violations are recorded.

Lead

The Mexican National Petroleum Company (PEMEX) has taken action to reduce the Pb content of petrol sold in the MCMA. Emission standards requiring the use of catalytic converters in new vehicles require such vehicles to use unleaded petrol. Current levels of atmospheric Pb in the MCMA are consistently near, or below, international norms. However, the current consensus is that no amount of Pb in the environment can be considered safe.

Ozone

Available data indicate that peak O_3 concentrations increased steadily between 1986 and 1992 but have declined steadily since then. As indicated in Table 6.8, the number of days on which the standard has been exceeded has

Table 6.8 Number of days exceeding the "Indece Metropolitano de la Calidad del Aire" (IMECA) values of 100, 200, 250 and 300 in Mexico

	Number of days exceeding IMECA values			
Year	IMECA of 100	IMECA of 200	IMECA of 250	IMECA of 300
1988	329	67	11	1
1989	329	15	3	0
1990	328	84	27	3
1991	353	173	56	8
1992	333	123	37	11
1993	324	80	14	1
1994	344	93	4	0
1995	324	88	6	0
1996	322	69	5	0

The Mexican Air Quality Standard for each pollutant corresponds to a value of 100

been unchanged over the past decade. One recent study which attempted to correct for meteorological variability indicated that, when corrected for meteorological factors, peak concentrations remained virtually unchanged and average values may still be increasing slightly (Cicero-Fernandez, 1995).

Suspended particulate matter

Both peak and annual-average SPM concentrations in Mexico City far exceed the Mexican health advisory level. The limited data on SPM composition suggest that the sources are primarily combustion and secondarily the central MCMA, while dust is also a major factor in outlying areas. Characterising the SPM and aerosols in Mexico City's air is one of the primary objectives of a recently initiated IMP Los Alamos study (see below). As a general rule, PM_{10} levels exceed the current air quality standard by approximately 30 per cent and have shown no downward or upward trends in recent years.

6.7.8 Progress to date

As a result of actions taken to date, significant improvements in air quality have now been documented compared with just a few years ago, especially in respect of Pb, CO and SO_2. However, as noted above, serious problems remain. While peak O_3 concentrations have been reduced, the number of days which exceed healthy levels appear unchanged and PM_{10} levels also remain very high.

Improvements in fuels have led to the most significant progress to date. Reductions in the Pb content of leaded petrol and the increasing sales of Magna Sin, the premium unleaded petrol, have been solely responsible for the improvement in ambient Pb concentrations noted above. A reduction in diesel fuel sulphur content to 0.05 per cent (weight) (the same levels as for the USA) has contributed to lower SO_2 concentrations. Other improvements in fuels, including lower volatility and aromatics content and the addition of oxygenates, have also been beneficial in lowering HC emissions. In turn they would tend to lower O_3 and toxic concentrations in the air.

The adoption of stringent standards for new vehicles, especially passenger cars, the replacement of older taxis, combis and micro buses with newer catalyst-equipped vehicles, and the conversion of cargo lorries to LPG fuel using certified, closed loop three-way catalysts, have also been very beneficial. In combination with an I/M programme, these strategies appear to have been responsible for the approximately 50 per cent improvement in CO concentrations in air.

6.7.9 Scientific base

A number of improvements in the scientific base for air quality control have occurred in recent years and this remains an important priority. A few of the most important improvements are described below.

The Departmento del Districto Federal (Cuidad de Mexico) (DDF) has funded a laboratory run by the University which, among other things, conducts independent tests of petrol quality. Since 1993, the Bacteriology and Physical Chemistry Laboratory of the DDF has conducted an independent petrol quality assurance programme. The acquisition of a chromatographic system using PIANO (paraffins, isoparaffins, aromatics, napthenes and olefins) software allows the laboratory to determine the percentages of more than 250 different hydrocarbons. About 20 samples are tested per week for aromatics, olefins, benzene and Pb, amongst other constituents.

Significant improvements have also occurred in the emissions inventory although all interested parties have agreed that additional improvements are necessary. Such improvements include:

- Analysis of traffic conditions in much more detail than in the past.
- Improved emission factors for vehicles based on actual test data from many more vehicles than in 1990.
- Much more careful analysis of area sources.
- Comprehensive studies of LPG leakage.

The NO_X inventory, however, is still considered to be very weak. In conjunction with the Los Alamos laboratory, an ozone model has been developed which most experts believe provides much greater insight to the O_3 problem. Based on this model, the role of NO_X control is now seen to be much more important than originally thought. In addition to promoting NO_X

control, the model indicates that substantial VOC control will be necessary if O_3 levels are to be lowered significantly.

Development of a second project, focused on SPM and aerosols, is underway between IMP and the Los Alamos laboratory . A recent series of health studies suggest that particulate matter concentrations may be the most serious air pollutant from a public health standpoint and that impaired visibility is also a serious concern. The major areas of focus in the study include monitoring and characterisation, health issues, visibility and the relative roles of primary and secondary pollutants. The study is expected to take four years and is still ongoing.

6.8 Los Angeles, USA

The Los Angeles (LA) Metropolitan Area is the main population centre of the California South Coast Air Basin which constitutes a 16,600 km^2 area bounded by the Pacific Ocean to the west and mountain ranges to the north and east. This topography, combined with the local weather patterns of light winds, sea breeze, subsidence inversions and high solar intensity, produce conditions of atmospheric stagnation conducive to pollutant reaction and build up. The Los Angeles Basin has been one of the most rapidly growing areas (in terms of population, economy and urban development) in the USA since the early 1940s.

The LA Basin is surrounded to the east by mountains with a high desert area beyond and has a very low annual rainfall. Sunshine occurs all year round and the winters are very mild. In the summer, a warm air cap often forms over the moist cool marine air layer and this inhibits vertical mixing. The dominant daily weather pattern is for an onshore breeze to begin in the morning after sunrise and for an offshore breeze to occur at night. During periods of stagnating high pressure, this circulation pattern, which takes pollutants out to sea at night and returns them to land during the day, allows air pollutants to build up in the air shed (tropospheric boundary layer) until a new weather front passes.

The LA Basin developed with almost no public transport network and, consequently, the residents have to rely on motor vehicles for almost all their transportation needs. The presence of large numbers of motor vehicles and their intensive consumer-orientated service activities, makes the Basin the most polluted area in the USA. The 8 million vehicles in an urban area of 12 million people represents possibly the greatest number of vehicles per person (0.67) in the world — almost everyone over 16 years old has a vehicle. Furthermore, the area is expected to grow by 5 million residents, with 68 per cent more vehicle-miles-travelled and 40 per cent more journeys, by the year 2010 (SCAG, 1990).

The air quality of LA has been a major concern since the early 1940s when the post-war boom brought rapid population growth and industrial expansion. Since 1947, the Los Angeles Air Pollution Control District (LAAPCD), the precursor of the South Coast Air Quality Management District (SCAQMD), began enforcing air pollutant emission controls. Hourly O_3 concentrations exceeding 0.6 ppm were reported, and in the 1960s O_3 frequently exceeded 0.5 ppm. In 1966, the State of California began instituting vehicle emission standards which were stricter than those promulgated later by the US EPA for the nation as a whole. Despite their strict controls, the maximum O_3 during the period 1986–91 only decreased to 0.35 ppm because of an 81 per cent population increase between 1960 and 1990 and because of allied increases in industrial activity and vehicular traffic. This enormous population increase resulted in many motorists commuting 60–80 miles each way to secure affordable single-family housing outside the central area.

The effects of motor vehicle air pollutants on human health have been investigated in several studies that have shown significant impacts on the health of residents in the LA area (Detels *et al.*, 1981; Hodgkin *et al.*, 1984; Euler *et al.*, 1988; Abbey *et al.*, 1990). It has been estimated that LA residents suffer O_3-related symptoms on 17 days each year and an increased mortality rate of 1 per 10,000 per year or 1,600 excess deaths per year related to PM_{10}. According to recent estimates, air pollution control to meet the national ambient air quality standards will cost US $ 10,000 million (Hall *et al.*, 1992).

Emissions of petrol-powered automobiles have been stringently controlled by the US EPA for the past 25 years, starting with vehicles manufactured in 1968. California has an exemption from the US Clean Air Act that allows it to have stricter emission standards than the rest of the country. The US EPA and the California Air Resources Board (CARB) standards are shown in Table 6.9 and it can be seen how lower values were often implemented by California several years before the US EPA. For example, the NO_x standard of 0.25 g km^{-1} which was adopted in California in 1992, was only adopted by the US EPA in 1995. The progressive tightening of petrol-powered automotive emissions is clearly illustrated in Table 6.9.

On 14 September 1990, CARB adopted vehicle emissions standards which would apply throughout the 1990s and into the next century. Specifically, a portion of new vehicles, starting in 1994, would have to meet an HC standard of 0.125 grams per mile when fuelled with conventional petrol (so-called transitional low emission vehicles or TLEVs) or be low emission vehicles (LEVs) emitting only 0.075 grams per mile of HC or be ultra low emission

Table 6.9 Comparison of US EPA and CARB emission standards, 1966–2003

	CO (g km^{-1})		HC (g km^{-1})		NO_x (g km^{-1})	
Year	US EPA	CARB	US EPA	CARB	US EPA	CARB
1966		32		3.70		3.1
1968	32	32	3.7	1.56	3.1	2.5
1970	24	15.6	2.1	1.56	3.1	2.5
1972	24	15.6	1.9	1.56	3.1	1.9
1973	24	15.6	1.9	1.56	1.9	1.9
1974	24	15.6	1.9	1.56	1.9	1.2
1975	9.3	5.6	0.9	0.44	1.9	1.2
1978	9.3	5.6	0.9	0.29	1.2	0.93
1980	5.6	5.6	0.25	0.29	1.2	0.46
1983	2.1	4.4	0.25	0.29	0.6	0.46
1992	2.1	4.4	0.25	0.29	0.6	0.25
1995	2.1		0.16		0.25	
2003	2.1		0.08		0.125	

Sources: CARB, 1992; US EPA, 1995

Table 6.10 50,000-mile certification standards for passenger cars operating on petrol in California

Category	NMOG (g/mile)	CO (g/mile)	NO_x (g/mile)
Adopted for 1993	0.25	3.4	0.4
TLEV	0.125	3.4	0.4
LEV	0.075	3.4	0.2
ULEV	0.040	1.7	0.2

NMOG Non-methane organic gases
LEV Low emission vehicles
TLEV Transitional low emission vehicles
ULEV Ultra low emission vehicles
Source: Walsh, 1996

vehicles (ULEVs) which emit no more than 0.04 grams per mile of HC and 0.2 grams per mile of NO_X. These standards are summarised in Table 6.10. Table 6.11 shows the different standards which could apply if vehicles are powered by alternative fuels. In addition, zero emission vehicles (ZEVs) must be introduced and must comprise at least 2 per cent of sales by the year 1998 and 10 per cent by 2003. In the aggregate, the vehicle combination must comply with average emissions levels listed in Tables 6.10–6.12.

In Table 6.10 volatile non-methane organic gases (NMOG) are substituted for conventional hydrocarbons because the constituents in the exhaust could

Table 6.11 NMOG emission standards in California for flexible- and dual-fuelled passenger cars when operating on alternative fuel and petrol at 50,000 miles

	NMOG emission standard (g/mile)	
Category	Alternative fuel, before reactivity adjustment	Petrol
TLEV	0.125	0.25
LEV	0.075	0.125
ULEV	0.040	0.075

NMOG Non-methane organic gases
LEV Low emission vehicles
TLEV Transitional low emission vehicles
ULEV Ultra low emission vehicles
Source: Walsh, 1996

Table 6.12 Implementation rates used to calculate fleet average standards for passenger cars in California

	Implementation rate (%) for emission standards (g/mile) of:						
Model year	0.39	0.25	0.125 TLEV	0.075 LEV	0.040 ULEV	0.00 ZEV	Fleet average standard (g/mile)
1994	10	80	10				0.250
1995		85	15				0.231
1996		80	20				0.225
1997		73		25	2		0.202
1998		48		48	2	2	0.157
1999		23		73	2	2	0.113
2000				96	2	2	0.073
2001				90	5	5	0.070
2002				85	10	5	0.068
2003				75	15	10	0.062

TLEV Transitional low emission vehicles
ULEV Ultra low emission vehicles
LEV Low emission vehicles
ZEV Zero emission vehicles
Source: Walsh, 1995

change as fuels change in the future; these emissions will be reactivity adjusted for cleaner burning fuels.

The percentage requirements for ZEVs in Table 6.12 are mandatory, indicating that the California programme includes a mandatory ZEV requirement. In the time frame under consideration, such vehicles will probably be battery electric or fuel cell electric. In response to the California mandate, a tremendous effort has begun to develop a commercially viable, electrically powered system. The General Motors Company has set up a new vehicle division and has made a commitment to introduce the "Impact" (an electrically powered car) in the mid 1990s. In addition, the domestic industry and the Department of Energy, amongst others, have put together a consortium focused on developing a more durable, longer range battery. While electric vehicles are not expected to have the same range as conventional vehicles in the short term, they appear quite viable for many commuter applications, such as urban delivery vehicles, etc.

The California LEV programme includes a banking and trading component which permits vehicle manufacturers to earn marketable credits and to make up for poor vehicle sales, or overly optimistic sales projections for a certain model year, by using credits previously earned by that manufacturer or a competitor. Credits earned in a given model year are discounted to 50 per cent of their original value if not used or sold by the end of the following year, and are further discounted to 25 per cent if not used within two model-years. Credits will have no value if not used by the beginning of the fourth model-year after being earned.

6.8.1 Air Quality Management Plan

In order to meet the California and USA air quality standards in the future, a three-tiered 1991 Air Quality Management Plan has been developed that will stipulate even more severe emission reductions than those given in Table 6.11 (SCAG, 1990). This plan is based on the September 1990 CARB Low-Emission Vehicles and Clean Fuels rules. The major elements of this strategy are:

- Reducing HC and NO_X emission standards by 80 per cent and 50 per cent respectively from the 1990 levels.
- Requiring the sale of ZEVs, starting in 1998.
- Allowing the use of vehicles powered by alternative fuels.
- Requiring availability of sufficient alternative fuels for such vehicles.

In addition, the plan calls for increasing the average number of passengers per vehicle from the present day 1.13 to 1.5 by 1999. This will be achieved by increasing funding for transit improvements and high occupancy vehicle (HOV) facilities, by using parking fees to discourage single passenger commuting and by providing facilities for commuting bicyclists. Alternative work weeks, telecommuting and employer "rideshare" incentives are also going to be introduced.

If the emission reductions are not sufficient to meet State and Federal air quality standards, then contingency actions may be necessary. These actions, including emission charges on petrol and diesel fuels, vehicle use and parking lots, time-and-place control measures and even limits on vehicle registrations, will be quite unpopular and may require special legislation.

6.9 References

Abbey, D.E., Euler, G.L., Hodgkin, J.E. and Magie, A.R. 1990 Long term ambient concentrations of total suspended particulates and oxidants as related to incidence of chronic disease in California Seventh-day Adventists. Abstract. Second Annual Meeting of the International Society for Environmental Epidemiology Berkeley, CA, 13–15 August 1990.

American Lung Association 1990 *The Health Costs of Air Pollution. A survey of studies published 1984–1989.* American Lung Association, Washington, D.C.

Bolade, T. 1993 Urban transport in Lagos. *The Urban Age*, **2**(3), 7–8.

Brosthaus, J., Waldeyer, H., Lacy Tomayo, R., Sanchez Martinez, S. 1994 *Air Pollution Control in the Mexico City Metropolitan Area (MCMA): Emissions Inventory for Mobile Sources — Tool for Scenario Calculations and Necessary Input for Dispersion Modelling.* Comision Metropolitana Para la Prevention y Control de la Contaminacion Ambiental en El Valle de Mexico, Mexico, D.F.

CARB 1992 *Emission Standards for Petrol-Powered Cars in California.* California Air Resources Board, Sacramento, CA.

Cicero-Fernandez, P. 1995 *Analisis Exploratorio De La Influencia Meteorologica En Las Tendencias Del Ozono En La Zona Metropolitana Del Valle De Mexico*. Comision Metropolitana Para la Prevention y Control de la Contaminacion Ambiental en El Valle de Mexico, Mexico, D.F.

Detels, R., Sayre, J.W., Coulson, A.H., Rokaw, S.N., Massey, F.J., Tashkin, D.P. and Wu, M.-M. 1981 The UCLA population studies of chronic obstructive respiratory disease. *American Review of Respiratory Disease*, **124**, 673–680.

Euler, G.L., Abbey, D.E., Hodgkin, J.E. and Magie A.R. 1988 Chronic obstructive pulmonary disease symptom effects of long-term cumulative exposure to ambient levels of total oxidants and nitrogen dioxide in California Seventh-day Adventist residents. *Archives of Environmental Health*, **43**, 279–285.

Fernandez-Bremauntz, A. 1992 Commuters exposure to carbon monoxide in the metropolitan area of Mexico City. PhD thesis. Centre for Environmental Technology, Imperial College of Science, Technology, and Medicine, University of London.

Fernandez-Bremauntz, A. and Merritt, J. Q. 1992 Survey of commuter habits in the metropolitan area of Mexico City. *Journal of Exposure Analysis and Environmental Epidemiology*, **2**, Suppl. 2, 95–112.

Hall, J.V., Winer, A.M., Kleinman, M.T., Lurmann, F.W., Brajer, V. and Colome, S.D. 1992 Valuing the health benefits of clean air. *Science*, **255**, 812–817.

Hodgkin, J.E., Abbey, D.E., Euler, G.L. and Magie, A.R. 1984 COPD prevalence in non-smokers in high and low photochemical air pollution areas. *Chest*, **86**, 830–838.

Hong Kong EPA 1989 Pollution in Hong Kong, a time to act. White Paper.

Khosa, M.M. 1993 Transport and the 'Taxi Mafia' in South Africa. *The Urban Age*, **2** (3), 8–9.

Mage, D. and Zali, O. [Eds] 1992 *Motor Vehicle Air Pollution. Public Health Impact and Control Measures*. WHO/PEP/92.4, World Health Organization and Republic and Canton of Geneva, Geneva.

National Environment Board of Thailand 1990 *Air and Noise Pollution in Thailand 1989*. National Environment Board of Thailand, Bangkok.

Raungchat, S. (Undated) Various factors associated with blood lead levels of traffic policemen in Bangkok Metropolis. Thesis.

SCAG 1990 Air Quality Management Plan. Draft. Southern California Association of Governments, South Coast Air Basin, South Coast Air Quality Management District, Los Angeles.

Shen, S-H. and Huang, K-H. 1989 Taiwan air pollution control programme: impact of and control strategies for transportation-induced air pollution. Paper presented at UN Conference, Ottawa, Canada.

Stickland, R. 1993 Bangkok's urban transport crisis. *The Urban Age*, **2** (1), 1–6.

TDRI, 1990 *Energy and Environment: Choosing The Right Mix*. The TDRI Year End Conference, Industrializing Thailand and its Impact On The Environment, Research Report No. 7, Thailand Development Research Institute, Bangkok.

Torres, E. and Subida, R. 1994 Impact of vehicular emissions on vulnerable population in Metro Manila, Abstract 191, ISEE/ISEA Joint Conference 1994, Program Abstracts, Research Triangle Park.

US EPA 1987 *Draft Regulatory Impact Analysis, Control of Gasoline Volatility and Evaporative Hydrocarbon Emissions From New Motor Vehicles*. Office of Mobile Sources, United Sates Environmental Protection Agency, Washington, D.C.

US EPA 1995 *Vehicle Emission Standards*. United Sates Environmental Protection Agency, Washington, D.C.

Walsh, M.P. 1994 *Carlines*, Issue 94-6, Arlington, VA.

Walsh, M.P. 1995 *Carlines*, Issue 95-6, Arlington, VA.

Walsh, M.P. 1996 *Carlines*, Issue 96-1, Arlington, VA.
Walsh, M.P. 1997 *Carlines*, Issue 97-4, Arlington, VA.
Walsh, M.P. 1998 *Carlines*, Issue 98-3, Arlington, VA.
WHO/UNEP 1992 *Urban Air Pollution in Megacities of the World*. Published on behalf of the World Health Organization and the United Nations Environment Programme by Blackwell Publishers, Oxford.

Chapter 7*

MOTOR VEHICLE POLLUTION CONTROL IN GENEVA

The Canton of Geneva is located in a basin formed by the mountains of neighbouring France (Figure 7.1). The area of the Canton is 246 km^2. Geneva first became established as a specific entity during the period of Roman domination. Its location as a junction for lake, road and river traffic, meant that it rapidly developed into a city. From the late sixteenth century onwards, Geneva became the starting point of many coaching routes for mail and passengers. However, it was not until the early nineteenth century that these activities began to extend beyond the neighbouring regions.

In 1814, the Republic and Canton of Geneva joined the Swiss Confederation. The demolition of the fortifications marked the transformation from a pre-industrial to an industrial city and ushered in a period of substantial urban expansion between 1850 and 1880. It also influenced the development of the road system because the present "inner ring" follows the line of the old city walls.

The present population is concentrated in the city of Geneva and in some surrounding communes where new settlements were built, mainly in the 1960s and 1970s (Figure 7.2). Some 80 per cent of the population live in an area representing less than a quarter of the Canton, but where 90 per cent of jobs are concentrated. In the communes of the Canton, other than the city itself, more than 50 per cent, and in some cases over 80 per cent of people commute to work (Primatesta, 1984). Local automobile traffic is heavy and is increased further by about another 28,000 cross-border commuters ("frontaliers") from the neighbouring districts of France (Ain and Haute-Savoie) and, according to the 1990 census, about another 18,000 inhabitants of the Canton of Vaud who also work in Geneva (OCS, 1996).

Since the completion of its motorway bypass in 1993, Geneva has become the centre of an important motorway network. A land-use zoning system aims to preserve the rural character of the Canton. Building land, including industrial estates and the airport, account for 31 per cent of the area of the Canton, leaving almost 55 per cent for agriculture and over 14 per cent for forests,

* *This chapter was prepared by F. Cupelin and O. Zali*

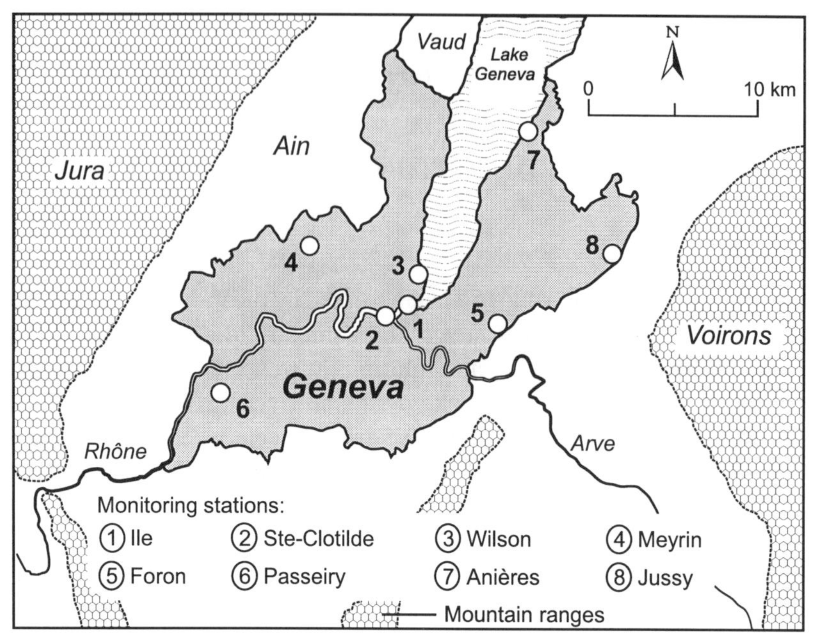

Figure 7.1 The Canton of Geneva, its borders, surrounding mountains and locations of the monitoring stations

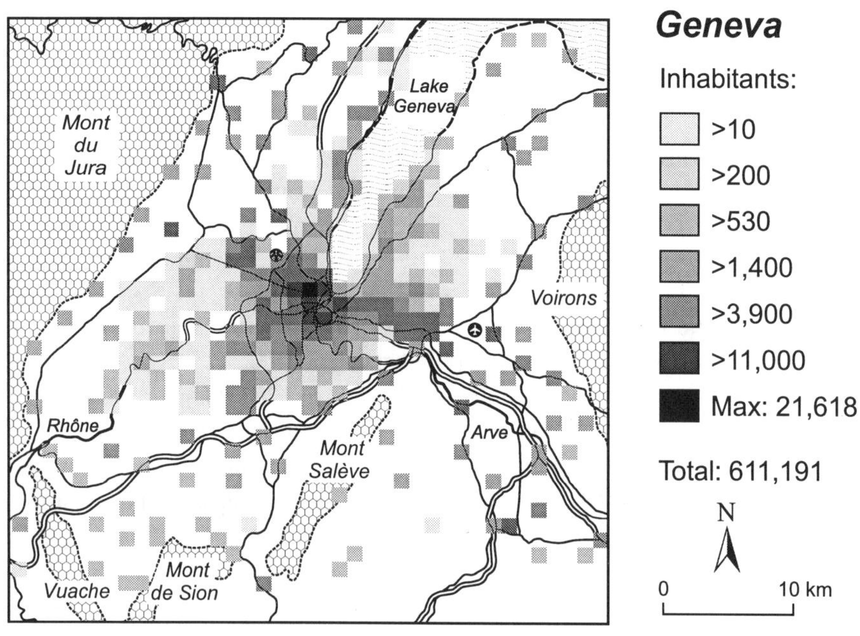

Figure 7.2 Repartition of the population in the Geneva region

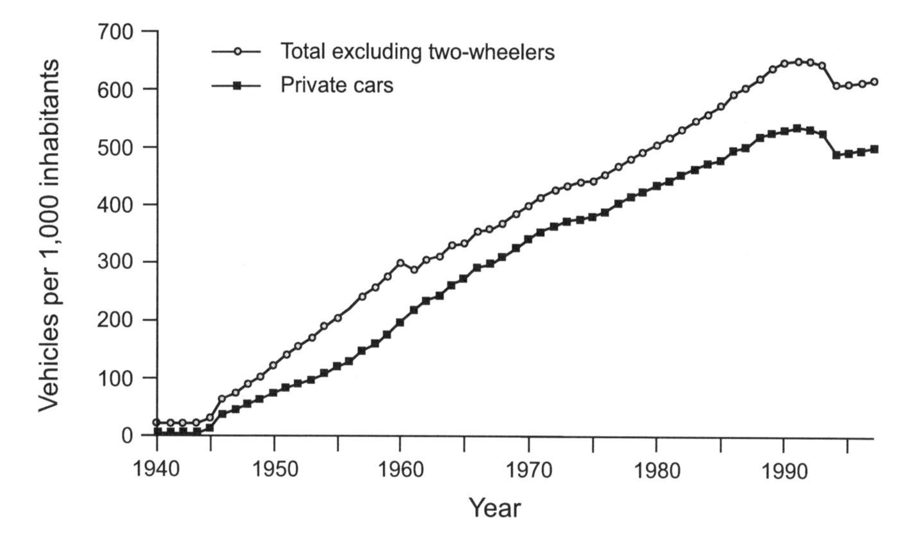

Figure 7.3 Total number of motor vehicles and number of cars per 1,000 inhabitants in the Canton of Geneva,1940–97 (Data from OCS, various years)

parks and recreational areas. Geneva is a major conference, congress and exhibition centre. It is also the home of the European Office of the United Nations, and a number of other UN specialised agencies also have their headquarters in the city.

7.1 Motor traffic and public transport

With 622 motor vehicles per 1,000 inhabitants, including 498 private cars (OCS, 1997), Geneva is one of the most highly motorised areas of the world (Figure 7.3). The parking problem is acute, mainly because of heavy commuter traffic. The number of access routes to Geneva is limited by its geographical location, especially by the presence of the lake. The Arve and Rhône rivers have to be crossed by bridges on which the traffic is extremely heavy. The Mont Blanc bridge is crossed by an average of 86,000 vehicles a day. The principle of a new lake crossing ("traversée de la rade") was approved by vote by the large majority of citizens in 1987. However, in 1996, under a very different economic climate, the majority voted to reject plans to build either a bridge or a tunnel. Traffic entering the city is regulated by an automatic system based on the counting of vehicles by means of induction coils. This allows the Office des Transport et de la Circulation (OTC) to adapt control systems to different conditions of traffic flow.

For a long time, car traffic was encouraged in the city. This led to the dismantling of the tram network which many people now regret. Geneva's first tram line was inaugurated in 1862. At the end of the last century there was a rapid growth in this form of transport together with the change from animal traction to steam-driven vehicles. At the beginning of twentieth century, 126 km of tram lines were electrified (Encyclopédie de Genève, 1984). With the increase in private cars, the tram lines were gradually taken out of use. On the urban and suburban routes the trams have been progressively replaced by trolley buses. The countryside is served by diesel-driven buses. However, one tram line remained in use throughout this period. The number of passengers transported daily increased by 10 per cent between 1989 and 1997. This increase was the result of a policy for developing demand which was initiated by Geneva Public Transport (Transports Publics Genevois (TPG)) with the support of the Cantonal authorities. As a reault, a second line was commissioned in 1995 and a third in 1998. Another project linking the two shores of the lake and extending to the French districts is also under study.

On the 6 km stretch of railway between Geneva Cornavin station and the new station serving Geneva airport, over 200 trains are circulating each day. This is evidence of the quality of the transport connections with the rest of Switzerland. Through the link with Bellegarde on the French rail network, Geneva also has a high-speed train service (Train Grande Vitesse (TGV)).

Geneva Airport, which handles about 100,000 flights and 6 million passengers each year, is another important component of the city's transport system.

7.2 The air pollution monitoring network

Continuous monitoring of air quality by fixed measuring stations began in 1972 at the Sainte-Clotilde station (Figure 7.1) which is located in the city centre very close to the ECOTOX (Service of Ecotoxicology) laboratories. The station at Anières is located in a rural area on the left shore of the lake and has been operational since 1973. In subsequent years, other stations were put into service at Meyrin in 1984 (a suburban environment fairly close to the airport), Jussy in 1985 (woodland environment), Ile in 1986 (urban environment, heavy traffic, town centre), Wilson in 1986 (urban environment, heavy traffic, lake side), Foron in 1987 (suburban environment, heavy traffic nearby) and, finally, Passeiry in 1989 (rural environment). As indicarted, all stations were set up at sites representative of a particular environment. The prevailing winds in Geneva are north-easterly (Bise) and south-westerly. The location of the stations at Anières and Passeiry were selected to meet the objective of monitoring and comparing air pollutant concentrations in the air masses moving along the main wind direction.

Air samples were taken on the roof of each station at a height of about 3 m, except at Jussy where a 16 m mast is used for taking measurements above the

Table 7.1 Air quality standards (valeurs limites d'immission) for Geneva stipulated by an Air Protection Order (OPair)

Substance	Limit value	Statistical definition
SO_2	30 µg m^{-3}	Annual mean (arithmetic mean)
	100 µg m^{-3}	95 percentile of half-hourly means over one year
	100 µg m^{-3}	24-hour average; should under no circumstances be exceeded more than once a year
NO_2	30 µg m^{-3}	Annual mean (arithmetic mean)
	100 µg m^{-3}	95 percentile of half-hourly means over one year
	80 µg m^{-3}	24-hour average; should under no circumstances be exceeded more than once a year
CO	8 mg m^{-3}	24-hour average; should under no circumstances be exceeded more than once a year
O_3	100 µg m^{-3}	98 percentile of half-hourly means over one month
	120 µg m^{-3}	hourly average; should under no circumstances be exceeded more than once a year
SPM [1]	70 µg m^{-3}	Annual mean (arithmetic mean)
	150 µg m^{-3}	95 percentile of 24-hour means over one year
Pb in SPM	1 µg m^{-3}	Annual mean (arithmetic mean)
Cd in SPM	10 ng m^{-3}	Annual mean (arithmetic mean)
Deposited PM (total)	200 mg m^{-2} d^{-1}	Annual mean (arithmetic mean)
Pb in deposited PM	100 µg m^{-2} d^{-1}	Annual mean (arithmetic mean)
Cd in SPM	2 µg m^{-2} d^{-1}	Annual mean (arithmetic mean)
Zn in deposited PM	400 µg m^{-2} d^{-1}	Annual mean (arithmetic mean)
Tl in deposited PM	2 µg m^{-2} d^{-1}	Annual mean (arithmetic mean)

PM Particulate matter
Source: Opair, 1986

[1] SPM is fine suspended particulate matter with terminal velocity < 10 cm s^{-1}

treetops. The atmospheric pollutants measured are SO_2, nitric oxide (NO), NO_2, CO, O_3, total HC and, at four of the eight stations, methane (CH_4). Concentrations of pollutants are measured continuously and every 30 minutes, a half-hourly average is taken. Five stations are equipped for recording wind speed and direction, and two have equipment for taking dust samples and for measuring sunlight.

The Air Protection Order (Ordonnance sur la Protection de l'Air, OPair) of 1 March 1986 set emission and air quality standards for Switzerland. Air quality standards are aimed at protecting human health and the environment from adverse effects caused by air pollution (Table 7.1). They also help to

Table 7.2 Number of times the daily ambient concentration limit values were exceeded for NO_2 and the number of times the hourly ambient concentration limit value was exceeded for O_3, 1987–97

	1987	1988	1989	1990	1991	1992	1993	1994	1995	1996	1997
Nitrogen dioxide											
Ste-Clotilde	38	17	63	49	37	38	18	12	6	2	3
Ile	121	75	103	135	99	62	43	28	26	8	18
Ozone											
Ste-Clotilde	166	91	661	400	731	284	160	142	60	195	34
Anières	783	758	1,100	765	1,169	827	427	487	377	411	203

Source: ECOTOX,1996

prevent annoyance within the general population. A distinction can be drawn between short-term standards (daily, hourly) intended to protect people from the effects of high concentrations of short duration, and long-term standards (annual) aimed at preventing chronic exposure.

A requirement to protect not only people but also the most sensitive forms of life has led the Swiss authorities to adopt very low air quality standards. These standards correspond, except in a few details, to the air quality guidelines recommended by WHO (WHO, 1987).

7.3 Trends in air quality in Geneva

Trends in annual mean concentrations of SO_2, CO, NO_2, and O_3 in air, together with the number of times air quality standards are exceeded each year, are the most important variables considered in the evaluation of air pollution in Geneva. The annual mean trends for SO_2, CO, NO_2 and O_3 at three urban monitoring sites are shown in Figure 7.4. A decreasing trend was also observed for CO, SO_2 and, since 1992, for NO_2. For O_3 the values remain approximately constant.

Short-term air quality standards (OPair Standards) are being attained for CO and SO_2 but not for O_3 and NO_2. In summertime, O_3 concentrations often exceed the hourly air quality standard, although the number of these exceedances and their magnitude is tending to decrease (Table 7.2).

The exceeding of the daily standard for NO_2 tends to be more pronounced in winter when stable anticyclonic situations cause the Geneva basin to be covered by a thick layer of stratus cloud. Consequently, pronounced temperature inversions prevent pollutants from dispersing.

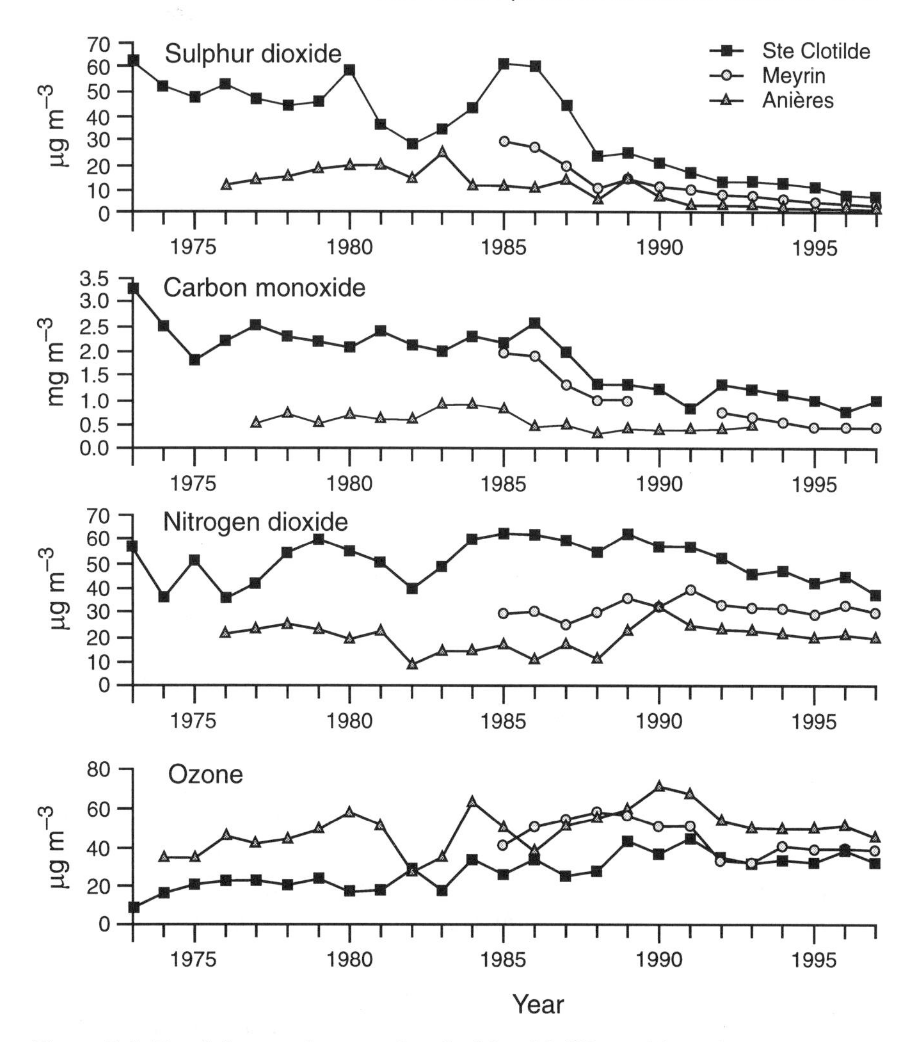

Figure 7.4 Trends in annual mean values for SO_2, CO, NO_2 and O_3 at three stations representative of urban areas (Ste-Clotilde), the suburbs (Meyrin) and rural areas (Anières)

Apart from the daily NO_2 standard of 80 $\mu g\ m^{-3}$ stipulated by the Federal Order, a specific cantonal regulation enforces a temporary reduction in traffic volume if the daily NO_2 mean exceeds 160 $\mu g\ m^{-3}$ on more than three consecutive days (République et Canton de Genève, 1989). In such an event, which has not yet occurred, cars without catalytic converters would only be allowed on the roads every other day.

7.4 The air quality management plan

7.4.1 Legal framework

Switzerland's federal environmental legislation considers the environment as a whole. Consequently, the impact of different kinds of pollution and subsequent corrective measures are dealt with individually, collectively and in relation to their combined effects. In situations where people or the environment have already suffered from adverse effects, corrective actions alleviate the problem quickly. Legislation must reconcile the development of human activities, especially future technology and the individual behaviour of consumers, with the burden that nature will have to sustain. The result of this is that the burden on the environment must not increase, in spite of demographic and economic growth, and it should even be reduced wherever possible.

In accordance with this principle, OPair has adopted the two-pronged approach of prevention and correction for pollution control. In the first prevention approach, atmospheric pollution is limited by enforcing compliance with emission standards. The costs of the necessary emission reduction devices is paid by the owner of the emission source, because Federal legislation recognises the "polluters pay" principle. Such emission control actions must be technically feasible and economically bearable because they are enforced on all sources irrespective of the prevailing air quality. Therefore the best available control technology for reducing air pollution should also be applied even when air pollutant concentrations are low. In the correction approach to pollution control, action is taken to implement additional control measures, including the possibility of limiting the use of private cars.

Where air quality standards are exceeded, Cantons must set up clean air implementation plans ("plans des mesures"). A deadline of 1 April 1994 was set for implementation. Difficulties encountered by Cantons when applying this new legislation have resulted in delays in preparing these plans.

Environment Impact Assessment (EIA) is a pollution prevention instrument. It allows an assessment to be made of the environmental impact of construction or alteration of a certain installation before any decision to build or alter is actually taken. Environment Impact Assessment can also be used to ensure that all the provisions of legislation on the environment are duly considered and it encourages the authors of action plans to improve them, if necessary. An EIA should be made during the planning of a project and, therefore, before the implementation of the project. It enables the promoter to recognise, and eventually to correct, conceptual errors and to review and revise the investment plan. It can, therefore, also serve as an instrument for economic management. The EIA report must be accessible to the public, allowing them to make their own judgement about the extent to which a project is compatible with environmental protection. In Geneva, various EIAs have already been performed, such as for the planning of a motorway bypass and for numerous parking area projects.

The OPair contains a large number of emission standards, applicable to existing and planned installations. Most of them are set for installations of a minimum size, or for maximum volumes of pollutant concentrations allowed in effluents. For large industrial installations and combustion plants, OPair makes a number of specific provisions. Boilers and burners above a certain capacity are subject to a standard expert evaluation prior to approval for sale. When an installation is found not to conform to the requirements of the OPair, it must be "cleaned up". A control action plan is then prepared in collaboration with the authorities which set a deadline for its completion. This is usually five years. Emissions from motor vehicles are limited by legislation on road traffic. Since 1986, all vehicles registered in Switzerland must comply with standards analogous to those in force in the USA since 1983. In 1995, new technical prescriptions were transcribed from the European legislation (OETV, 1995; OETV1, 1995).

7.4.2 Strategy

Geneva developed a clean air implementation plan in 1989 and, after consultation with the main sectors concerned, the plan was adopted by the Conseil d'Etat of the Canton of Geneva on 27 March, 1991.

High NO_2 and O_3 concentrations prevailed in Geneva. Ozone is a secondary pollutant produced by the joint activity of NO_X, VOC and sunlight. The O_3 problem is not limited to the Geneva region and has to be treated on a larger scale. This section concentrates only on the strategy and tactics used to reduce total NO_X emissions. Ozone can be reduced by the reduction of primary pollutants and therefore a clean air implementation plan with respect to NO_X would also affect ambient O_3 concentrations. Future developments of the Geneva plan will ensure parallel reduction in VOC emissions. In 1999, an incentive tax will be introduced which will progress from CHF 1 per kilogram to CHF 3 per kilogram in 2008.

In Switzerland, the Cantons are responsible for checking the quality of air for their territory. The results are assessed in relation to OPair objectives. If air pollutant concentrations are excessive, the cantonal authority must prepare a clean air implementation plan to prevent or eliminate excessive air quality concentrations. The plan must indicate:

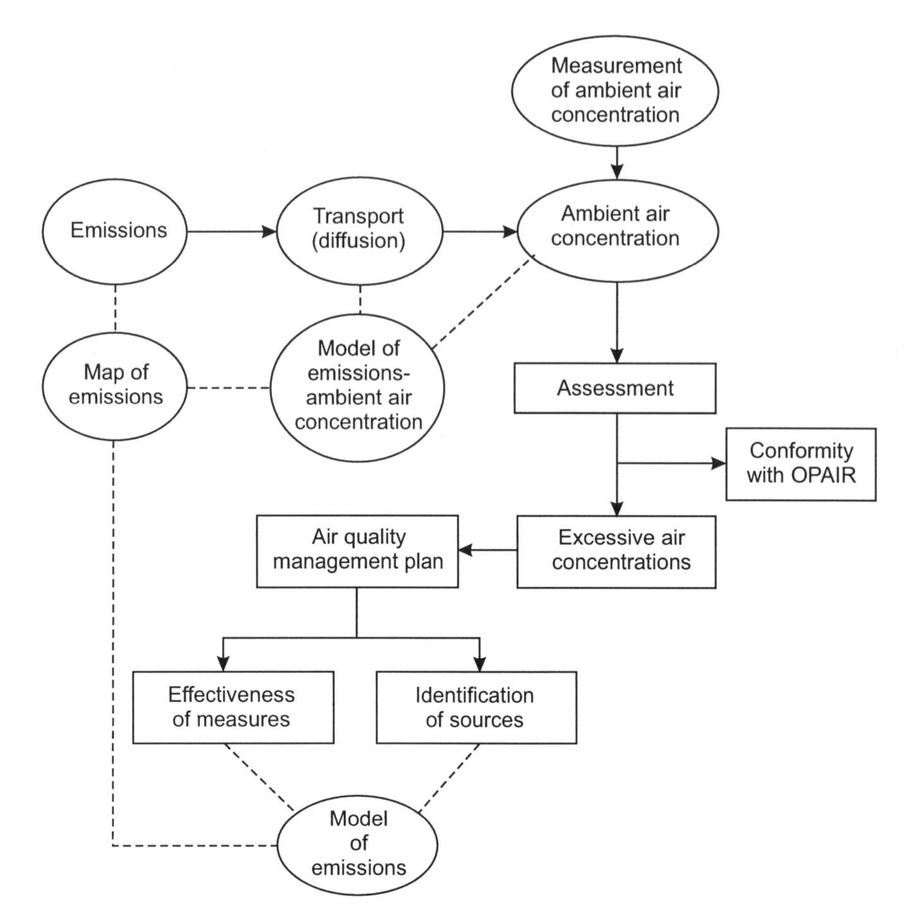

Figure 7.5 Methodology for the development of the air quality management plan in Geneva

- The sources of emissions responsible for excessive ambient air pollutant concentrations and the proportion of the total burden of pollution for which they are responsible.
- Control measures to be taken at the sources and an estimate of the effectiveness of each of those measures.

Figure 7.5 shows the methodological procedure for implementing the clean air implementation plan. As a rule, the control measures listed in the plans are implemented within five years.

Several instruments are required for the development of a clean air implementation plan. These include emissions inventories, inventories of air pollutant concentrations monitored and or simulated, numerical models for

Table 7.3 Specifications for the numerical model used to assess emission sources in Geneva

Questions	Specifications
Who is producing the emissions?	Qualitative inventory of sources
How much is emitted?	Estimate of emissions from each source
Where are the sources of emissions?	Map of sources
How will these emissions develop?	Evaluation of measures set out in the plan

estimating emissions and air pollutant concentrations (Table 7.3), emission and air quality standards, models for the causal assignment of the most polluting sources and models for assessing the effectiveness of control actions.

The method used to estimate ambient air pollutant concentrations from emissions is based on the results obtained from the model of emissions and uses a dispersion model. It is essential to ensure that the control measures taken enable the air quality standards to be attained. The application of a dispersion model to simulate ambient air pollutant concentrations is based on the assumption that there is a relationship between emissions and ambient air pollutant concentrations. Fluctuations in emissions are one of the causes of fluctuations in air pollutant concentrations. Other important factors are changes in the meteorological parameters responsible for dispersion, such as wind speed and direction and atmospheric stability. In order to take these factors into account, the model is based on measurements made within the Geneva Network for Observation of Atmospheric Pollution (Reseau d'Observation de la Pollution Atmosphérique à Genève (ROPAG)). The measurements have been correlated with data on estimated emissions around the measuring station.

7.4.3 Emissions inventory for NO_X

The formation of NO_X results mainly from reactions between atmospheric oxygen and nitrogen during combustion processes. A small proportion arises from oxidation of nitrogen-containing compounds in the combustion materials themselves. Total nitrogen oxides are always considered when evaluating emissions. Total nitrogen oxides includes a large number of compounds of the chemical composition N_yO_X although, with respect to air protection, the only significant compounds are NO and NO_2 (denoted together as NO_X). Nitric oxide, which accounts for 90–95 per cent of NO_X emissions, is ultimately changed in the atmosphere into the more toxic NO_2. For this reason,

NO_X emissions are expressed in terms of mass of NO_2, even if only 5–10 per cent of the NO_X emitted is actually NO_2.

There are three main sources of emissions of total NO_X: transport, heating and industry. In Geneva, road traffic is a major source of air pollution because of the tremendous growth in motor traffic over the past 30 years. Geneva is one of the most motorised areas in the world.

Aircraft emissions from the Geneva International Airport are another source of NO_X. For the territory of the Canton of Geneva, only emissions on the ground and during takeoff and landing have been considered in the clean air implementation plan. Regionally, only NO_X emitted into the atmosphere up to a height of 800 m is of real importance. Emissions in the upper troposphere may be assumed to have no influence on local air pollutant concentrations.

In addition to transportation and industry, NO_X emissions from heating systems are also significant. There are two sources of emissions from heating systems:

- Homes, where emission levels are linked to the number of inhabitants; and
- Workplaces, where emissions are linked to the number of employees irrespective of the type of activity.

Emissions from domestic and industrial heating systems, unlike road traffic emissions, show considerable seasonal variation. However, when such emissions are compared with traffic emissions on an annual basis, this fact is not evident.

In the Canton of Geneva, approximately 75 per cent of the working population is in the administrative sector. Therefore, NO_X emissions produced by industry are relatively low and result from incineration of domestic waste, industrial power plants, civil engineering activities, and agriculture and forestry. Nitrogen oxides from industrial plants are emitted to the atmosphere through chimneys. The height of these chimneys is chosen to ensure that there is no excessive air pollution from a single source. Emissions from civil engineering, agriculture and forestry activities come from mobile sources. It is therefore difficult to estimate their contribution to an emission inventory.

In order to estimate NO_X emissions some simplifying hypotheses are required. Appendix 7.1 describes how these estimates have been made. The model takes into account two types of emission. The first type includes emissions from vehicle traffic and heating and has been mapped in 500×500 m^2 grid cells. Emissions produced by aircraft moving on the ground at the airport are included in the traffic emissions. The second type comprises diffuse sources of emissions, i.e. industrial emissions emitted through high chimneys and emissions from air traffic up to 800 m above ground. These emissions increase NO_X concentrations over the whole territory and are not linked

Table 7.4 Emissions of NO_X for the Canton of Geneva, 1988

Source	NO_X ($t\ a^{-1}$)
Road traffic	4,093
Air traffic	556
Heating installations	1,008
Industries	910
Total	**6,567**

Source: Service de la législation et des publications officielle, 1991

directly to given grid squares. Table 7.4 summarises the emissions of total NO_X for Geneva in 1988.

7.4.4 Inventory of ambient air pollutant concentrations

An ideal model for the estimation of ambient air pollutant concentrations should permit calculation of the half-hourly concentration of pollutants and their statistical means using emission data, physical–chemical and photochemical parameters of the atmosphere, and climatic and meteorological conditions. Such a model could be used in conjunction with weather forecasts to predict air pollution conditions. At present, such models are relatively complicated for an urban environment with many sources of air pollutants and they also require many different data. Therefore, a simpler, empirical multilinear regression model for the calculation of air pollutant concentrations has been applied in the Canton of Geneva (Cupelin *et al.,* 1995). This model is based on the assumption of a linear relationship between emissions and ambient air pollutant concentrations. The model was calibrated using data supplied by the ROPAG network.

Estimates of future ambient air pollutant concentrations will be expressed in average annual concentrations. Incorporating the types of meteorological conditions and their annual frequency of occurrences is fairly straightforward. Emissions for traffic and heating are also calculated for annual periods. In this way, the NO_X annual mean forecast can be compared with the annual mean air quality standard of OPair.

The locations of the monitoring sites and their special features were described in section 7.2. It is important to establish whether the monitoring stations are representative of the whole territory of Geneva. Pollutant sources in the vicinity of the site could make the monitoring station representative of only a small area around the pollutant source and thereby distort the interpretation of general air pollutant concentrations. In theory, the influence of a single source depends on the distance between the source of emission and the monitoring site. Experience shows that this is true for primary pollutants, such as SO_2, but that it does not apply to secondary pollutants resulting from physical–chemical transformations such as O_3 and NO_2. For the latter, maximum ambient air pollutant concentrations vary with meteorological conditions and solar intensity. Therefore, total emissions in a grid square of 1 km^2, centred at the monitoring site, have been taken into account. This procedure provides a good correlation between emissions and NO_2 concentrations.

Ambient concentrations at a given location depend on the quantity of pollutant emitted and on the conditions of dispersion. A two-step method, which allows the influence of weather conditions on air pollution to be taken into account has been developed. In the first step, the weather conditions for each day are designated according to a classification system published by the Institut Suisse de Météorologie (ISM). In the second step, these standardised meteorological situations are grouped together as a function of the dispersion conditions and the transport of pollutants. For Geneva, there are six typical classes.

Using the air pollutant concentration data for a reference year, the parameters of the model for different groups of typical meteorological situations are set. In this manner it is possible to calculate, for any year, the mean value of ambient air concentrations of NO_2 using the frequency of occurrence of the different categories and the emissions for a reference year.

In Geneva, the mean monthly NO_2 concentrations were found to be relatively constant whereas, in the urban area, the emission inventory shows that a large proportion of NO_x emissions were produced by heating installations (even though these were operational only in winter). Total NO_x emissions should, therefore, be related more effectively to average NO_x concentrations than to average NO_2 concentrations. The model for establishing this relationship is presented in Appendix 7.1. Figure 7.6 shows the inventory of NO_2 concentrations in air for 1988, calculated from the emissions inventory. The air pollutant concentration values obtained for each grid square were interpolated to provide a map where NO_2 concentrations are shown with isometric lines.

7.4.5 Control of air pollution

Evaluation of total NO_2 emissions showed that road traffic was the main source. Thus the section of the clean air implementation plan devoted to control measures for road traffic would be most important. The progressive replacement of vehicles without catalytic converters by new ones conforming to the USA 1983 standards gradually leads to a major reduction in NO_x emissions. However, this is not sufficient to reach OPair objectives or to resolve the problems associated with the saturation of the Canton road network.

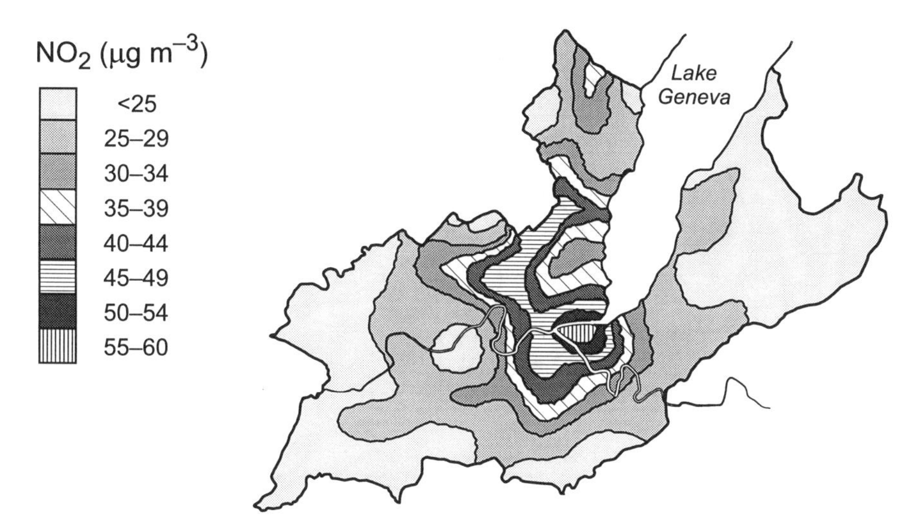

Figure 7.6 Map of average NO_2 concentrations for the Canton of Geneva, 1988, calculated from the emissions inventory (Figure courtesy of ECOTOX)

Measures related to transport

All the control actions described here were developed by the cantonal transport office and are described in full in the clean air implementation plan (Publications Officielles, 1991). All the control measures proposed follow five guiding principles which were established with political consensus, namely:

- Compliance with federal orders governing environmental protection.
- An impact that complements other measures leading to overall improvement of the environment.
- Being respected and, therefore, having the support of the population.
- Preservation of individual liberty in terms of mobility and choice.
- Fostering development of the viability of the city and canton.

Some of the control measures proposed have been put into practice immediately, while others depend on completion of public works or transport infrastructure and are being phased-in gradually.

Regular information on the clean air implementation plan is distributed to the general public so that every inhabitant feels involved. The reduction of emissions from road traffic is an ambitious project requiring the support of the population. It is particularly ambitious because its aims appear, at first sight, contradictory, such as reducing road traffic emissions while maintaining or even increasing the mobility of the population. To meet these aims, it is necessary to transfer some of the use of private vehicles to other, less

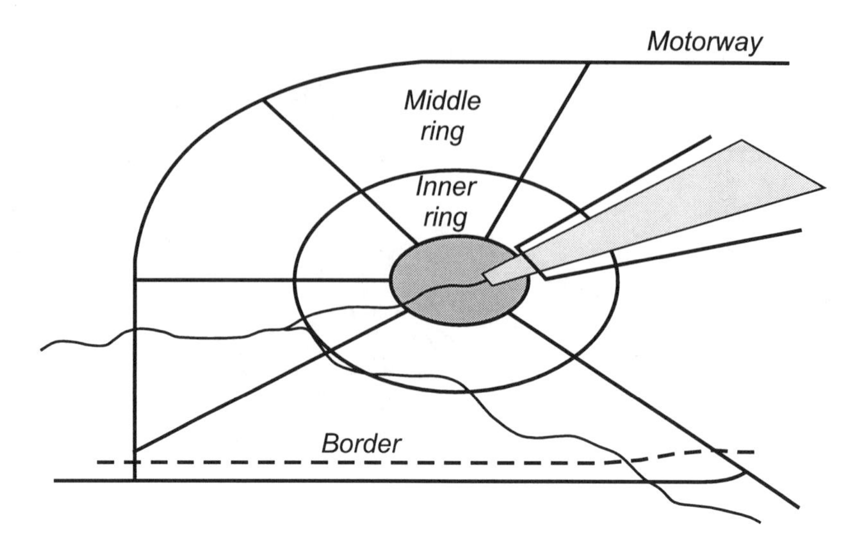

Figure 7.7 Radial/concentric traffic plan for Geneva (After Office des transports et de la circulation, Département de justice et police, Geneva)

polluting, forms of transport. The public transport network was intended to meet this need and therefore its development was a priority.

A number of steps were taken to improve the performance of the existing public transport network in terms of capacity, availability, speed and comfort. The maximum available level was governed by the peak hour public transport needs. A master plan for the public transport network for the period 1990–94 was adopted which included improvement of existing lines, provision of preferential traffic lights at cross-roads, development of the urban network with the creation of a further tram line, and reorganisation of the Geneva regional network. The master plan for the public transport network of 1995–2000 aims to create a new structure for the urban network (trams and underground trains) and to develop regional transport (trains and buses). Taken together, these measures will encourage a larger proportion of journeys to be made in public rather than in private transport.

The inventory of road traffic emissions showed that the centre of the town was the area where emissions were highest, leading to excess air pollutant concentrations. In order to reduce these, the city centre traffic has to be reduced, by reducing the ring-based traffic system (Figure 7.7) with a system based on sectors (Figure 7.8). In the sector system, passage from one sector to a neighbouring sector is possible along a link road, whereas passage across two sectors is not possible through the centre. Passage through two non-adjacent sectors is only possible using a road outside the sectors. This new

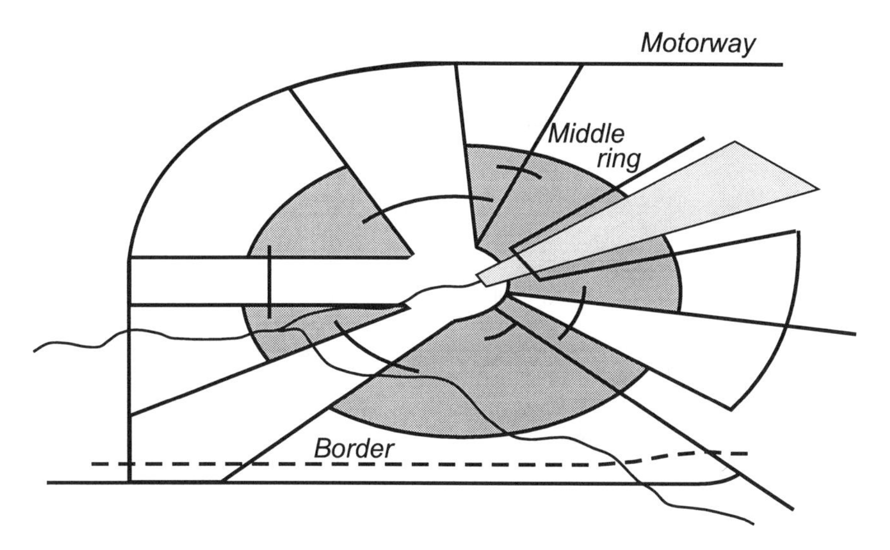

Figure 7.8 Sector based traffic plan for Geneva (After Office des transports et de la circulation, Département de justice et police, Geneva)

traffic system maintains access to the centre, but reduces traffic passing through it by creating sealed areas that are accessible only to terminal traffic (OTC 1990). The system has been introduced gradually after the bypass motorway was opened in 1993. It should progressively reduce traffic in transit through the town.

There has been concern that new arrangements would lead to a displacement of industry and housing towards the new routes, creating unexpected traffic problems. However this should be avoided if the regulations preventing changes in land-use zoning are applied strictly. Therefore, the new traffic plan objectives must be backed up with appropriate land-use policies.

The existing traffic management system is being modernised and updated to make the best use of existing and planned infrastructure. The flow of public transport vehicles in the town is a priority feature of the traffic management system. A "by-pass" has been developed to improve the priority for public transport by allowing public vehicles to pass queues and move almost as though in a traffic-free area. Cantonal regulations are being adopted to ensure the coexistence of the new form of traffic network with the supply of parking capabilities. The control of parking regulates the traffic for regular, visiting and professional drivers.

The establishment of "park and ride" car parks at the points of entry to the city are being encouraged. The building of underground car parks enables surface parking spaces to be removed and the road space can be redistributed

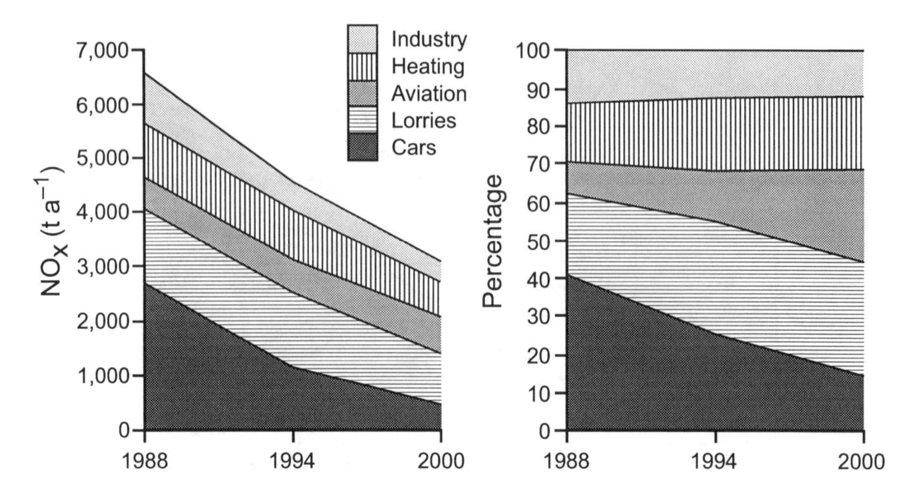

Figure 7.9 Sources of NO_2 emission in the Canton of Geneva as absolute values and as relative proportions

for the benefit of public transport, deliveries, two-wheeled vehicles, pedestrians and green spots. Present and planned parking spaces for regular commuters are being counted and, where possible, reallocated to the inhabitants of the districts and to visitors.

Measures related to air traffic

The Canton of Geneva enjoys a Federal concession for the use of its airport. Therefore, the influence of the Canton of Geneva with regard to air pollution from aircraft is rather limited. The Federal authorities have been asked to consider, in the interests of pollution control, restricting access to Swiss airports for certain types of aircraft which are deemed too "dirty". Landing surcharges could be used for aircraft that cause too much pollution.

Measures related to heating and industry

Control measures for the domestic and industrial sectors have also been developed and evaluated in the Geneva clean air implementation plan (Publications Officielles, 1991). Figure 7.9 provides the absolute and relative total NO_X emissions from transport and domestic and industrial sources. The latter sources are beyond the scope of this book and are not discussed further.

7.4.6 Evaluation of control measures

Control measures have been evaluated in two ways. In the first approach, the proposed measures were evaluated in relation to emissions. This involved a

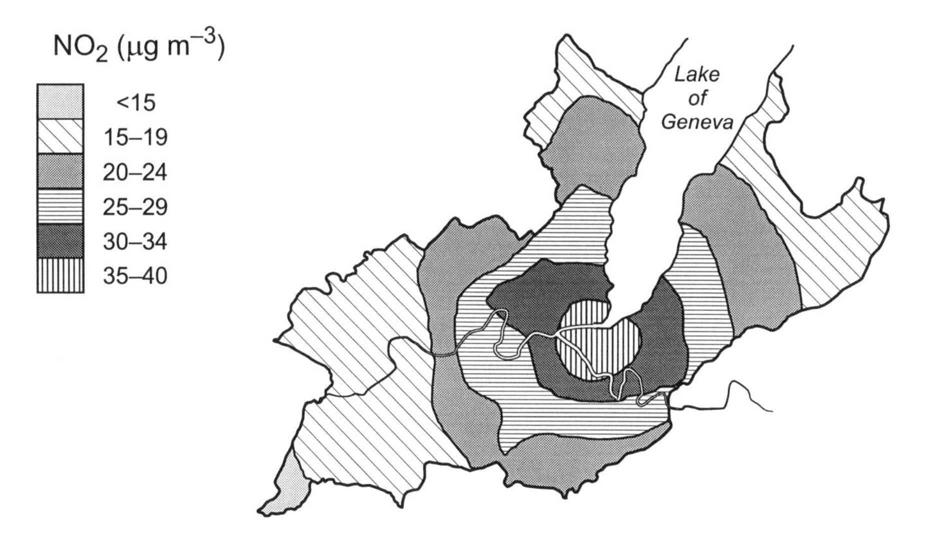

Figure 7.10 Map of average NO_2 concenrations for the Canton of Geneva in the year 2003, as predicted by the model

sectoral approach to the various listed sources of total NO_x. In the second approach, the proposed model was used to calculate future air pollutant concentrations. Emissions will be reduced step by step. Two schedules for future air pollutant concentrations were selected, one for 1994, which is when the OPair ended, and the other for the year 2000, by which time all the proposed measures should have been carried out.

Evaluation of control measures to reduce NO_2 emissions of road traffic
According to the OTC, the set of measures proposed should eventually reduce private motor traffic by 15 per cent overall, with reductions of up to 40 per cent in some parts of the city centre and up to 20 per cent in some rural areas. By contrast, the new traffic plan would lead to more traffic on the by-pass motorway. With this linear reduction in road traffic, overall emissions calculated on the basis of these hypotheses should have been 2,496 t a^{-1} of NO_x for 1994 and should be 1,367 t a^{-1} of NO_x for the year 2000. These calculations took account of the gradual introduction of the catalytic converter, which should further reduce emission factors.

According to the model, the background concentration of NO_2 would fall at the same time as total NO_x emissions. Figure 7.10 shows the distribution of NO_2 concentrations in the year 2003 according to the model. They should be within the limit value of 30 $\mu g\ m^{-3}$ except in the inner-city, where more stringent measures will have to be taken if necessary.

7.5 Conclusions

The clean air implementation plan will enable the achievement of the objectives of the legislation on environmental protection and will minimise health risks for the population. Therefore, it will not give instant results and its application needs to be supervised rigorously for years to come. Every two years, revised estimates of the emissions are undertaken. Comparison of current ambient concentrations with estimated trends allow possible reinforcement of the proposed measures. In 1996 the "Rapport du Suivi 1995" was published (DIER, 1996). Up to the present time, the clean air implementation plan had only taken into account the reduction of NO_X, although it also deals with the problem of VOC. The measures have been analysed in terms of their effectiveness in reducing air pollution, although no cost–benefit analyses have yet been undertaken. The plan will be successful only if the majority of the population agree with its objectives. Therefore, an information campaign has to be planned, in parallel with the progressive introduction of the various control measures discussed in the clean air implementation plan.

7.6 References

Cupelin, F., Landry, J.C.L., Lehmann, P. and Despot, F. 1995 Modèle statistique pour la gestion de la qualité de l'air. *Science of the Total Environment*, **169,** 45–52.

DIER 1996 *Rapport du suivi du plan des mesures, Etat 1995*, Département de l'Intérieur, de l'Environnement et des Affaires Régionales, Genève.

ECOTOX 1998 *Mesure de la qualité de l'air à Genève.* Service of Ecotoxicology, Genève.

Encyclopédie de Genève, 1984 *Tome 3: La vie Des Affaires*. Genève.

ISM, 1996, *Classes HBP, 1p.,* Institut Suisse de Météorologie, Zurich, Suisse.

OCS (various years) *Annuaires statistiques du canton de Genève*, Office Cantonal de Statistique, Genève.

OETV 1995 *Ordonnance Concernant les Exigences Techniques Requises pour les Véhicules Routiers*. Berne.

OETV1 1995 *Ordonnance Concernant les Exigences Techniques Requises pour les Voitures Automobiles de Transport et Leurs Remorque*. Berne.

Opair 1986 *Ordonnance sur la Protection de l'Air*. Berne

OTC 1990 *Données de Base du Système de Circulation Genevois*. Office des Transports et de la Circulation , Genève.

Primatesta, G. 1984 *Paysages Genevois.* Delachaux et Nestlé [Ed.] Neuchâtel-Paris.

République et Canton de Genève, 1989 *Règlement (H 1 1,3) relatif à la restriction temporaire de la circulation motorisée en cas de pollution de l'air*, Genève.

Publications Officielles, 1991 *Assainissement de l'air à Genève, plan de mesures au sens de l'article 31 de l'OPair*, Genève.

WHO 1987 *Air Quality Guidelines for Europe*, World Health Organization, Regional Office for Europe, Copenhagen.

Appendix 7.1

ESTIMATION OF NO_X EMISSIONS IN GENEVA

Estimation of emissions from road transport

Estimation of annual emissions of total NO_X from road traffic requires knowledge of the following parameters:

Parameter	Definition
Volume of traffic	Vehicle flow, daily distribution
Type of vehicle	Private cars, delivery vehicles, lorries, buses, motorcycles, mopeds
Vehicle speed	Depending on type of road and observance of speed limits
Emission coefficient	Type of driving, mechanical state of vehicles, year of manufacture

The precision of the overall estimate therefore depends on the accuracy with which the above parameters are quantified. For the purpose of this estimation it was necessary to establish simplifying hypotheses for each of the parameters.

The NO_X emissions resulting from road transportation were calculated by Dériaz (1993) using a transport model which is based on a simulated 90-minute peak period in the morning during which travel from home to work or home to school are predominant. The "EMME/2" software incorporates a module which enables the calculation of traffic emissions for each link and for each grid cell, in this case 500 × 500 m. The new traffic management plan (OTC, 1992), of which the first elements were implemented in 1993, has been taken into account in the traffic flow model.

The emission factors which were used are those that were established by the Office federal de la protection de l'environnement des forêts et du paysage (OFEFP) (OFEFP, 1988) for Swiss vehicles and by the INRETS (Joumard and Lambert, 1991) for the French commuter portion of the traffic. The influence of the speed of the vehicles on the emissions was calculated using the following relationship:

$$F = \sum_{n=1}^{n=4} a_0 + a_n \cdot v^n$$

where: F = emission factor

a = parameters

v = speed determined by the traffic model

Table 7a Emissions per category of vehicle in relation to total number of vehicles in 1988

Vehicle category	% of total	Number	Emission factor ($g\ km^{-1}$)[1]	Relative emission
Passenger cars	79	188,497	1.5	65.5
Heavy goods vehicles	5	10,906	13.35	33.8
Motorcycles	7	16,820	0.11	0.4
Mopeds	9	21,331[2]	0.06	0.3

[1] For 1988, speed was taken at 50 km h^{-1}

[2] According to the Geneva automobile department, 1990

Types of vehicles

Road traffic consists of several categories of vehicles. In order to make an estimate of the proportion of emissions attributable to each vehicle category, their respective emissions were calculated on the basis of the known number of vehicles of each category registered in Geneva. Table 7a gives the relative emissions per category. The main categories of vehicles emitting NO_X were private cars and heavy lorries. As a result, the emissions map calculated on the basis of the loading, was prepared by taking only these two categories of vehicles into account. By counting vehicle movements, Dériaz (1993) established the average proportion of heavy lorry traffic in different zones of the canton. The proportion of heavy lorries was very important because the total NO_X emission factor for a lorry is approximately 10 times that of private car.

Model for NO_2 emission estimates

The basic assumption of the model is that a relationship exists between NO_X emissions and concentrations.

The influence of the rest of the built-up area was taken into account by introducing an additional variable as a function of the distance from the city centre (*D*). This approach seemed appropriate because the emissions map revealed that the main source was relatively homogenous and perpendicular to the prevailing winds.

The model allowed the calculation of the annual average NO_X concentration from averages of NO_X for a given meteorological category of type *i*, by making use of the following relationships:

$$[NO_x]_i = A_i + B_i \cdot E + C_i \frac{E_{tot}(year)}{E_{tot}(1988)} e^{-D/D_c}$$

where: $[NO_x]_i$ = annual average total NO_X for meteorological category i
E = total emission in a 1 km^2 grid cell
E_{tot} = total emissions in the canton
D = distance of the measuring station from the city centre
A, B, C, D_c = parameters

The parameters A, B, C and K were determined by multiple regression for each meteorological category, i, from the NO_X emission data and 1988 and 1991 pollution measurements. The coefficient D_C was determined from emission and concentration data for the years 1988 and 1991.

Finally, the annual averages of NO_X and NO_2 were calculated taking into account the percentage occurrence of the different categories during the year under consideration. The concentration of NO_2 is calculated from the expression:

$$[NO_2]_i = K_i\,[NO_x]_i^{\,0.616}$$

where: $[NO_2]_i$ and $[NO_x]_i$ = average annual concentrations for meteorological category i
K_i = parameter

Estimation of emissions caused by heating units

Methodology

A significant portion of the total NO_X emissions was linked to the use of fossil fuels for heating and hot water production. The evaluation of the total NO_X emissions is an arduous task which depended on the following parameters for each installation:

- Type of fuel.
- Fuel consumption.
- Concentration of total NO_X in the exhaust gases.

For each 500 × 500 m element of the grid, the emissions were calculated according to the OFEFP method (OFEFP, 1987a), based on the number of jobs, the number of inhabitants and the emission coefficients. Degree-days of heating provided (defined as the sum of the daily differences between the temperature of heated buildings (20 °C) and the average daily temperature) was equal to, or below, 12 °C (OCS, various years).

Emission coefficients for heating

A distinction was made between the energy necessary to heat dwellings and that needed for heating workplaces. It was assumed that 90 per cent of energy was consumed in the form of gas or extra-light fuel oil, 3 per cent for wood and coal, and the remainder as mainly electricity. This produced an emission coefficient of 45 kg of NO_x per tonne of fuel. The number of degree-days for the reference year of 1988 was 3,004. The following energy factors were used:

- 8.22 MJ a^{-1} per resident and per degree-day
- 17.54 MJ a^{-1} per workplace and per degree-day

The emission coefficients for total NO_x per resident and per workplace were obtained in the following manner:

- 1.1 kg a^{-1} NO_x per resident
- 2.4 kg a^{-1} NO_x per workplace

Estimation of emissions due to industry

Geneva is equipped with a large municipal solid waste incinerator. As a result of a project to expand its capacity, an environmental impact assessment was published in 1983 (Département des travaux publics, Geneva). The total NO_x emissions from this source for 1988 were estimated at 600 tonnes.

Some industrial heating installations used heavy fuel oil. A campaign to measure the emissions from these installations showed that they were responsible for approximately 65 per cent of the 200 tonnes emitted by industrial heating units. Emissions from industry not related to heating were estimated by the OFEFP method (OFEFP, 1987b). For the canton of Geneva, they were in the order of 110 t a^{-1}.

Control actions to be taken in relation to heating

New and more binding standards governing air protection came into force on 1 February 1992. These were expected to result in a 40 per cent reduction of total NO_x emissions between 1992 and 2010.

In order to reduce polluting emissions due to heating, energy-saving measures are being introduced: including the systematic checking of the installation of thermostats, provision of adequate regulation of domestic heating appliances and individual billing of heating and hot water costs.

Control actions for industry

The proportion of total NO_x attributable to industry was low. Furthermore, stricter limitations on emissions came into force on 1 February 1992 with the purpose, as mentioned above, of reducing the emissions by about 40 per cent between 1992 and 2010. The incinerator for domestic waste from the Canton, which is the source of a large proportion of industrial emissions, was extended in 1995 and thus emissions of NO_x have increased by 56 per cent. In order to limit the impact, an emission limit value of 80 mg m^{-3} was selected. This means that total NO_x emissions should fall from 940 to 150 t a^{-1} of NO_x

when the system to reduce NO_X is actually in place, which is planned for the year 2000.

According to Confederation estimates, introduction of the new OPair requirements in 1992 would lead to a 43 per cent reduction in emissions from heating installations. The predicted effect will not be fully apparent until the year 2000. The estimated change for that reference year would be 575 tonnes of NO_X per year. For industrial sites, the reduction in emissions would be similar to that in heating installations. For the incineration plant of the Canton of Geneva, the OPair standards of 1992 would not be met until the year 2000, at which point total NO_X emissions for the industrial sector would be 373 tonnes of NO_X per year.

References

Département des travaux publics, 1988 *Rapport d'Impact sur l'Environnement; Adaptation des Installations Cantonales de Traitement des Résidus de Cheneviers III*. Département des travaux publics, Geneva.

Dériaz, B. 1993 *Emissions d'Oxydes d'Azote du Trafic Routier*. Rapport 187, ECOTOX, Geneva.

Joumard, R. and Lambert, J. 1991 *Evaluation des Émissions de Polluants par les Transports en France de 1970 à 2010*. Rapport INRETS No 143, Bron, France.

OCS various years *Annuaires Statistiques du Canton de Genève*, Office Cantonal de Statistique, Geneva.

OFEFP 1987a *Les Cahiers de l'Environnement No 73; Comment Établir un Cadastre d'Émission*. Office fédéral de la protection de l'environnement, des forêts et du paysage.

OFEFP 1987b *Les cahiers de l'environnement No 76; émissions polluantes en Suisse dues à l'activité humaine (de 1950 à 2010)*. Office fédéral de la protection de l'environnement, des forêts et du paysage.

OFEFP 1988 *Les Cahiers de l'Environnement No 55; Émissions Polluantes du Trafic Routier Privé de 1950 à 2000*. (1986, supplément 1988) Office fédéral de la protection de l'environnement, des forêts et du paysage.

OTC 1992 *Conception Globale de la Circulation à Genève, Circulation 2000*. Office des transports et de la circulation, Geneva.

Chapter 8*

CONCLUSIONS AND RECOMMENDATIONS

Problems associated with motor vehicle air pollution and noise are increasing with the growth of motor vehicle traffic in developed and developing countries. The important air pollutants with respect to health effects include SPM (including Pb), CO, NO_2, HC, O_3 and SO_2. The conclusions and recommendations presented here arise from the detailed discussion given in the preceding chapters and are intended to provide guidance for the formulation and implementation of effective policies to prevent serious air pollution problems from occurring or worsening in countries at all stages of economic development.

The most serious air pollution problem in the world today is caused by motor vehicles in metropolitan areas. The metropolitan areas of the world are inhabited by some 50 per cent of the earth's population and these areas either already have, or soon will have, motor vehicle traffic congestion problems that will lead to high concentrations of certain air pollutants. Although serious air pollution problems can be caused locally by industrial emissions, and regionally or seasonally by combustion of high sulphur content fossil fuels (i.e. coal and oil), the influence of emissions from the large and growing motor vehicle fleets is becoming more and more predominant.

Motor vehicles and their associated emissions seriously damage the health of urban populations. Suspended particulate matter, whether emitted in diesel exhaust or formed as aerosols by atmospheric photochemical reactions, can cause pulmonary irritation, contribute to the exacerbation of respiratory illnesses, and increase total and respiratory disease-related mortality. Lead is a pernicious component of SPM and can create a severe neurophysiological health hazard, especially for children living near areas of high traffic density. In urban areas, CO is almost entirely emitted by vehicular traffic and affects people with cardiovascular deficiencies. Nitrogen dioxide, which is mostly produced from nitrogen monoxide, causes a decrease in atmospheric visibility, changes the oxidation behaviour of the atmosphere, and can act as a

* *This chapter was prepared by Dietrich Schwela*

respiratory irritant. Hydrocarbons, such as benzene or benzo(a)pyrene, which are found in all exhaust emissions are carcinogenic. Ozone, formed by the complex photochemical reactions of NO_X and HC, causes eye irritation, contributes to pulmonary irritation, provokes asthmatic attacks in susceptible individuals, and contributes to the development of chronic obstructive pulmonary diseases in repeatedly exposed populations. Sulphur dioxide emitted mostly in diesel exhaust can exacerbate respiratory diseases. Motor vehicle noise creates a constant disturbance to urban life and provokes annoyance reactions. It is also possible that there may be synergistic effects arising from the accumulation of air pollutants and noise.

8.1 Recommendations

1. Personal monitoring
It has been established that ambient air quality measured at fixed monitoring stations, designed to represent community exposure in general, underestimates significantly exposure to primary motor vehicle air pollutants (i.e. SPM, CO, HC and Pb) of many population subgroups living close to traffic-congested streets. People in the vehicles in heavy traffic, people walking and working along busy streets, and people whose homes front onto busy streets, are all exposed to air pollutants at much higher concentrations than those reported by community monitoring networks. These high exposures have been the subject of special studies of short duration in only a few locations. More personal monitoring is needed in order to assess the real exposure of the people.

2. Information management
Human exposure to high pollution levels in cities may be greater in developing than in industrialised countries because, in many developing counties vehicle emissions are higher and some lifestyles place people in close proximity to roadways. More information is needed on the exposure to automotive air pollutants of the general public and of high exposure population subgroups. From the few exposure studies undertaken in recent years in cities with little or no motor vehicle emission controls in place, evidence shows that drivers, commuters and street-side groups are exposed to extremely high concentrations of SPM, CO and Pb. There is, currently, little reliable data on community exposures, but there is an urgent need for the initiation of monitoring campaigns to establish the magnitude of this problem.

The primary data necessary to assess the magnitude of vehicular emissions in an urban area include the types of vehicles (i.e. passenger cars, lorries, motorcycles) their age, the number, the speed and the daily kilometres driven.

This information must then be combined with emission factors by category to derive the automotive emissions for a given urban area. Whereas the primary set of data is usually obtainable in developed and developing countries, emission factors have to be calculated from test drive procedures and are often only available in developed countries. Care must be taken in applying emission factors derived from one location to different locations, without correcting for local vehicle character, fuel composition and driving patterns. The magnitude of emissions may be grossly underestimated if assessments in developing countries are based on emission factors that apply to test conditions in developed countries. Account must also be taken of the fact that emission factors are established in standardised conditions and actual emissions from in-use vehicles are increased significantly by age, poor maintenance and wear-and-tear.

3. Emission controls, and inspection and maintenance

Emission controls aimed directly at vehicle engines can limit automotive emissions by introducing changes in engine design, mandatory I/M and the use of catalytic converters to reduce the pollutants emitted. A different fuel composition can also reduce emissions. Special low emission and ultra-low emission cars can be produced, but they present direct costs to the vehicle user. The possibility of using renewable energy sources should be pursued aggressively.

4. Changes in human behaviour

Changes in human behaviour and changes in traffic patterns are necessary to reduce the total volume of vehicular traffic and, thereby, to reduce emissions. Improving driving behaviour (such as by driving more smoothly, avoiding unnecessary acceleration and breaking) will also help to reduce emissions. Indirect controls on drivers, such as increasing fuel and vehicle prices, limiting urban parking, encouraging car pools, and providing low-fare mass transit, can reduce the number of vehicles on the road as well as the number of kilometres driven.

5. Present investment for long-term benefits

Some developed countries have made significant reductions in vehicular emissions. Countries such as Switzerland, Japan, the USA and the member states of the European Union have successfully introduced direct and indirect controls on vehicle emissions so that further air quality deterioration has not occurred. In most cases air quality has improved significantly. If these emission controls had not been imposed, air quality would have deteriorated so badly that the public would have incurred massive health costs. A high initial capital investment is required in order to reap these much greater, long-term benefits.

Vehicle emission controls require such heavy investment, but they should be seen in the perspective of long-term cost benefits to public health. Countries with motor vehicle pollution problems should begin to phase-in emission controls appropriate to their economic and social context. Sophisticated control measures can be implemented in countries where the financial resources are available. Countries with fewer resources can reduce emissions progressively by first instituting the simplest and most cost effective controls, such as I/M programmes and the use of cleaner fuels. This should be followed by a phased reduction of emissions by more costly measures which meet public health needs and are affordable.

6. Active prevention and plans for the future

Even the least developed countries should plan now to prevent motor vehicle air pollution problems from occurring in the future when significant economic development starts to occur. In the next century, as the world population increases further and economic development accelerates, the per capita number of vehicles in use in developing countries will increase at a staggering rate. Unless control measures are applied to reduce the use of vehicles, many growing urban areas will experience the same problems as other large cities do today, such as Bangkok, Mexico City or Los Angeles. Urban planners and environmental agencies should co-operate today and plan future transportation such that it includes mass transit, phased emission controls and I/M programmes. In this way, many countries could prevent the problems from becoming unmanageable in the future.

7. World-wide exchange of information and experience

Through the sharing of information and experience on a world-wide basis, countries will be able to choose motor vehicle emission control programmes which meet their needs most appropriately. By examining the history of motor vehicle air pollution control in both developed and developing countries, it will be possible to evaluate the success and failure patterns of the various direct and indirect control procedures that have been attempted, and to choose the control option sequence that is appropriate for each specific situation.

The newly developed WHO Air Management Information System (AMIS) under the umbrella of the Healthy Cities Programme is an example of successful information exchange on air quality management within the framework of a Global Air Quality Partnership.

Index